SSAT UPPER LEVEL MATH FOR BEGINNERS

The Ultimate Step by Step Guide to Preparing for the SSAT Upper Level Math Test

By

Reza Nazari

About Effortless Math Education

Effortless Math Education operates the www.effortlessmath.com website, which prepares and publishes Test prep and Mathematics learning resources. Effortless Math authors' team strives to prepare and publish the best quality Mathematics learning resources to make learning Math easier for all. We Help Students Learn to Love Mathematics.

All inquiries should be addressed to:
info@effortlessMath.com
www.EffortlessMath.com

ISBN: 978-1-64612-953-9

Published by: **Effortless Math Education Inc**

for Online Math Practice Visit www.EffortlessMath.com

Welcome to
SSAT Upper Level Math Prep
2021

Thank you for choosing Effortless Math for your SSAT Upper Level Math test preparation and congratulations on making the decision to take the SSAT Upper Level test! It's a remarkable move you are taking, one that shouldn't be diminished in any capacity. That's why you need to use every tool possible to ensure you succeed on the test with the highest possible score, and this extensive study guide is one such tool.

If math has never been a strong subject for you, **don't worry**! This book will help you prepare for (and even ACE) the SSAT Upper Level test's math section. As test day draws nearer, effective preparation becomes increasingly more important. Thankfully, you have this comprehensive study guide to help you get ready for the test. With this guide, you can feel confident that you will be more than ready for the SSAT Upper Level Math test when the time comes.

First and foremost, it is important to note that this book is a study guide and not a textbook. It is best read from cover to cover. Every lesson of this "self-guided math book" was carefully developed to ensure that you are making the most effective use of your time while preparing for the test. This up-to-date guide reflects the 2021 test guidelines and will put you on the right track to hone your math skills, overcome exam anxiety, and boost your confidence, so that you can have your best to succeed on the SSAT Upper Level Math test.

This study guide will:

☑ Explain the format of the SSAT Upper Level Math test.

☑ Describe specific test-taking strategies that you can use on the test.

☑ Provide SSAT Upper Level Math test-taking tips.

☑ Review all SSAT Upper Level Math concepts and topics you will be tested on.

☑ Help you identify the areas in which you need to concentrate your study time.

☑ Offer exercises that help you develop the basic math skills you will learn in each section.

☑ Give **2 realistic and full-length practice tests** (featuring new question types) with detailed answers to help you measure your exam readiness and build confidence.

This resource contains everything you will ever need to succeed on the SSAT Upper Level Math test. You'll get in-depth instructions on every math topic as well as tips and techniques on how to answer each question type. You'll also get plenty of practice questions to boost your test-taking confidence.

In addition, in the following pages you'll find:

➢ **How to Use This Book Effectively** – This section provides you with step-by-step instructions on how to get the most out of this comprehensive study guide.

➢ **How to study for the SSAT Upper Level Math Test** – A six-step study program has been developed to help you make the best use of this book and prepare for your SSAT

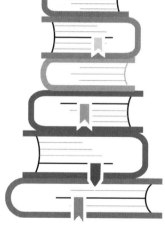

Upper Level Math test. Here you'll find tips and strategies to guide your study program and help you understand SSAT Upper Level Math and how to ace the test.

➢ **SSAT Upper Level Math Review** – Learn everything you need to know about the SSAT Upper Level Math test.

➢ **SSAT Upper Level Math Test-Taking Strategies** – Learn how to effectively put these recommended test-taking techniques into use for improving your SSAT Upper Level Math score.

➢ **Test Day Tips** – Review these tips to make sure you will do your best when the big day comes.

Effortless Math's SSAT Upper Level Online Center

Effortless Math Online SSAT Upper Level Center offers a complete study program, including the following:

✓ Step-by-step instructions on how to prepare for the SSAT Math test

✓ Numerous SSAT Math worksheets to help you measure your math skills

✓ Complete list of SSAT Math formulas

✓ Video lessons for all SSAT Math topics

✓ Full-length SSAT Math practice tests

✓ And much more...

No Registration Required.

Visit effortlessmath.com/SSAT to find your online SSAT Upper Level Math resources.

How to Use This Book Effectively

Look no further when you need a study guide to improve your math skills to succeed on the math portion of the SSAT Upper Level test. Each chapter of this comprehensive guide to the SSAT Upper Level Math will provide you with the knowledge, tools, and understanding needed for every topic covered on the test.

It's imperative that you understand each topic before moving onto another one, as that's the way to guarantee your success. Each topic provides you with examples and a step-by-step guide of every concept to better understand the content that will be on the test. To get the best possible results from this book:

➢ **Begin studying long before your test date**. This provides you ample time to learn the different math concepts. The earlier you begin studying for the test, the sharper your skills will be. Do not procrastinate! Provide yourself with plenty of time to learn the concepts and feel comfortable that you understand them when your test date arrives.

➢ **Practice consistently**. Study SSAT Upper Level Math concepts at least 20 to 30 minutes a day. Remember, slow and steady wins the race, which can be applied to preparing for the SSAT Upper Level Math test. Instead of cramming to tackle everything at once, be patient and learn the math topics in short bursts.

➢ Whenever you get a math problem wrong, **mark it off, and review it later** to make sure you understand the concept.

➢ Start each session by **looking over the previous material.**

➢ Once you've reviewed the book's lessons, **take the practice test at the back of the book** to gauge your level of readiness. Then, review your results. Read detailed answers and solutions for each question you missed.

➢ **Take another practice test** to get an idea of how ready you are to take the actual exam. Taking the practice tests will give you the confidence you need on test day. Simulate the SSAT Upper Level testing environment by sitting in a quiet room free from distraction. Make sure to clock yourself with a timer.

How to Study for the SSAT Upper Level Math Test

Studying for the SSAT Upper Level Math test can be a really daunting and boring task. What's the best way to go about it? Is there a certain study method that works better than others? Well, studying for the SSAT Upper Level Math can be done effectively. The following six-step program has been designed to make preparing for the SSAT Upper Level Math test more efficient and less overwhelming.

Step 1 - Create a study plan
Step 2 - Choose your study resources
Step 3 - Review, Learn, Practice
Step 4 - Learn and practice test-taking strategies
Step 5 - Learn the SSAT Upper Level Test format and take practice tests
Step 6 - Analyze your performance

STEP 1: Create a Study Plan

It's always easier to get things done when you have a plan. Creating a study plan for the SSAT Upper Level Math test can help you to stay on track with your studies. It's important to sit down and prepare a study plan with what works with your life, work, and any other obligations you may have. Devote enough time each day to studying.

It's also a great idea to break down each section of the exam into blocks and study one concept at a time.

It's important to understand that there is no "right" way to create a study plan. Your study plan will be personalized based on your specific needs and learning style.

Follow these guidelines to create an effective study plan for your SSAT Upper Level Math test:

★ **Analyze your learning style and study habits** – Everyone has a different learning style. It is essential to embrace your individuality and the unique way you learn. Think about what works and what doesn't work for you. Do you prefer SSAT Upper Level Math prep books or a combination of textbooks and video lessons? Does it work better for you if you study every night for thirty minutes or is it more effective to study in the morning before going to work?

★ **Evaluate your schedule** – Review your current schedule and find out how much time you can consistently devote to SSAT Math study.

★ **Develop a schedule** – Now it's time to add your study schedule to your calendar like any other obligation. Schedule time for study, practice, and review. Plan out which topic you will study on which day to ensure that you're devoting enough time to each concept. Develop a study plan that is mindful, realistic, and flexible.

★ **Stick to your schedule** – A study plan is only effective when it is followed consistently. You should try to develop a study plan that you can follow for the length of your study program.

★ **Evaluate your study plan and adjust as needed** – Sometimes you need to adjust your plan when you have new commitments. Check in with yourself regularly to make sure that you're not falling behind in your study plan. Remember, the most important thing is sticking to your plan. Your study plan is all about helping you be more productive. If you find that your study plan is not as effective as you want, don't get discouraged. It's okay to make changes as you figure out what works best for you.

STEP 2: Choose Your Study Resources

There are numerous textbooks and online resources available for the SSAT Upper Level Math test, and it may not be clear where to begin. Don't worry! This study guide provides everything you need to fully prepare for your SSAT Upper Level Math test. In addition to the book content, you can also use Effortless Math's online resources. (video lessons, worksheets, formulas, etc.) On each page, there is a link (and a QR code) to an online webpage which provides a comprehensive review of the topic, step-by-step instruction, video tutorial, and numerous examples and exercises to help you fully understand the concept.

You can also visit EffortlessMath.com/SSAT to find your online SSAT Upper Level Math resources.

STEP 3: Review, Learn, Practice

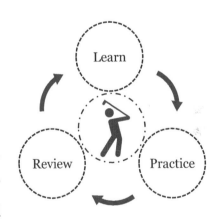

This SSAT Upper Level Math study guide breaks down each subject into specific skills or content areas. For instance, the percent concept is separated into different topics–percent calculation, percent increase and decrease, percent problems, etc. Use this study guide and Effortless Math online SSAT Upper Level center to help you go over all key math concepts and topics on the SSAT Upper Level Math test.

As you read each topic, take notes or highlight the concepts you would like to go over again in the future. If you're unfamiliar with a topic or something is difficult for you, use the link (or the QR code) at the bottom of the page to find the webpage that provides more instruction about that topic. For each math topic, plenty of instructions, step-by-step guides, and examples are provided to ensure you get a good grasp of the material.

Quickly review the topics you do understand to get a brush-up of the material. Be sure to do the practice questions provided at the end of every chapter to measure your understanding of the concepts.

STEP 4: Learn and Practice Test-taking Strategies

In the following sections, you will find important test-taking strategies and tips that can help you earn extra points. You'll learn how to think strategically and when to guess if you don't know the answer to a question. Using SSAT Upper Level Math test-taking strategies and tips can help you raise your score and do well on the test. Apply test taking strategies on the practice tests to help you boost your confidence.

STEP 5: Learn the SSAT Upper Level Test Format and Take Practice Tests

The SSAT Upper Level *Test Review* section provides information about the structure of the SSAT Upper Level test. Read this section to learn more about the SSAT Upper Level test structure, different test sections, the number of questions in each section, and the section time limits. When you have a prior understanding of the test format and different types of SSAT Upper Level Math questions, you'll feel more confident when you take the actual exam.

Once you have read through the instructions and lessons and feel like you are ready to go – take advantage of both of the full-length SSAT Upper Level Math practice tests available in this study guide. Use the practice tests to sharpen your skills and build confidence.

The SSAT Upper Level Math practice tests offered at the end of the book are formatted similarly to the actual SSAT Upper Level Math test. When you take each practice test, try to simulate actual testing conditions. To take the practice tests, sit in a quiet space, time yourself, and work through as many of the questions as time allows. The practice tests are followed by detailed answer explanations to help you find your weak areas, learn from your mistakes, and raise your SSAT Upper Level Math score.

STEP 6: Analyze Your Performance

After taking the practice tests, look over the answer keys and explanations to learn which questions you answered correctly and which you did not. Never be discouraged if you make a few mistakes. See them as a learning opportunity. This will highlight your strengths and weaknesses.

You can use the results to determine if you need additional practice or if you are ready to take the actual SSAT Upper Level Math test.

Looking for more?

Visit effortlessmath.com/SSAT to find hundreds of SSAT Upper Level Math worksheets, video tutorials, practice tests, SSAT Upper Level Math formulas, and much more.

Or scan this QR code.

No Registration Required.

SSAT Upper Level Test Review

The SSAT, or Secondary School Admissions Test, is a standardized test to help determine admission to private elementary, middle and high schools.

There are currently three Levels of the SSAT:

- ✓ Upper Level (for students in 3rd and 4th grade)
- ✓ Middle Level (for students in 5th-7th grade)
- ✓ Upper Level (for students in 8th-11th grade)

There are six sections on the SSAT Upper Level Test:

- ✓ Writing: 25 minutes.
- ✓ Math section: 25 questions, 30 minutes
- ✓ Reading section: 40 questions, 40 minutes
- ✓ Verbal section: 60 questions, 30 minutes
- ✓ Math section: 25 questions, 30 minutes
- ✓ Experimental: 16 questions, 15 minutes.

In this book, there are 2 complete SSAT Upper Level Math Practice Tests. Take these tests to see what score you'll be able to receive on a real SSAT Upper Level test.

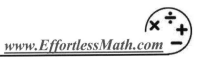

SSAT Upper Level Math Test-Taking Strategies

Here are some test-taking strategies that you can use to maximize your performance and results on the SSAT Upper Level Math test.

#1: USE THIS APPROACH TO ANSWER EVERY SSAT UPPER LEVEL MATH QUESTION

- Review the question to identify keywords and important information.

- Translate the keywords into math operations so you can solve the problem.

- Review the answer choices. What are the differences between answer choices?

- Draw or label a diagram if needed.

- Try to find patterns.

- Find the right method to answer the question. Use straightforward math, plug in numbers, or test the answer choices (backsolving).

- Double-check your work.

#2: USE EDUCATED GUESSING

This approach is applicable to the problems you understand to some degree but cannot solve using straightforward math. In such cases, try to filter out as many answer choices as possible before picking an answer. In cases where you don't have a clue about what a certain problem entails, don't waste any time trying to eliminate answer choices. Just choose one randomly before moving onto the next question.

As you can ascertain, direct solutions are the most optimal approach. Carefully read through the question, determine what the solution is using the math you have learned before, then coordinate the answer with one of the choices available to you. Are you stumped? Make your best guess, then move on.

Don't leave any fields empty! Even if you're unable to work out a problem, strive to answer it. Take a guess if you have to. You will not lose points by getting an answer wrong, though you may gain a point by getting it correct!

#3 : Ballpark

A ballpark answer is a rough approximation. When we become overwhelmed by calculations and figures, we end up making silly mistakes. A decimal that is moved by one unit can change an answer from right to wrong, regardless of the number of steps that you went through to get it. That's where ballparking can play a big part.

If you think you know what the correct answer may be (even if it's just a ballpark answer), you'll usually have the ability to eliminate a couple of choices. While answer choices are usually based on the average student error and/or values that are closely tied, you will still be able to weed out choices that are way far afield. Try to find answers that aren't in the proverbial ballpark when you're looking for a wrong answer on a multiple-choice question. This is an optimal approach to eliminating answers to a problem.

#4 : Backsolving

All questions on the SSAT Upper Level Math test will be in multiple-choice format. Many test-takers prefer multiple-choice questions, as at least the answer is right there. You'll typically have five answers to pick from. You simply need to figure out which one is correct. Usually, the best way to go about doing so is "backsolving."

As mentioned earlier, direct solutions are the most optimal approach to answering a question. Carefully read through a problem, calculate a solution, then correspond the answer with one of the choices displayed in front of you. If you can't calculate a solution, your next best approach involves "backsolving."

When backsolving a problem, contrast one of your answer options against the problem you are asked, then see which of them is most relevant. More often than not, answer choices are listed in ascending or descending order. In such cases, try out the choices B or C. If it's not correct, you can go either down or up from there.

#5 : PLUGGING IN NUMBERS

"Plugging in numbers" is a strategy that can be applied to a wide range of different math problems on the SSAT Upper Level Math test. This approach is typically used to simplify a challenging question so that it is more understandable. By using the strategy carefully, you can find the answer without too much trouble.

The concept is fairly straightforward–replace unknown variables in a problem with certain values. When selecting a number, consider the following:

- Choose a number that's basic (just not too basic). Generally, you should avoid choosing 1 (or even 0). A decent choice is 2.

- Try not to choose a number that is displayed in the problem.

- Make sure you keep your numbers different if you need to choose at least two of them.

- More often than not, choosing numbers merely lets you filter out some of your answer choices. As such, don't just go with the first choice that gives you the right answer.

- If several answers seem correct, then you'll need to choose another value and try again. This time, though, you'll just need to check choices that haven't been eliminated yet.

- If your question contains fractions, then a potential right answer may involve either an LCD (least common denominator) or an LCD multiple.

- 100 is the number you should choose when you are dealing with problems involving percentages.

SSAT Upper Level Math – Test Day Tips

After practicing and reviewing all the math concepts you've been taught, and taking some SSAT Upper Level mathematics practice tests, you'll be prepared for test day. Consider the following tips to be extra-ready come test time.

Before Your Test

What to do the night before:

- **Relax!** One day before your test, study lightly or skip studying altogether. You shouldn't attempt to learn something new, either. There are plenty of reasons why studying the evening before a big test can work against you. Put it this way–a marathoner wouldn't go out for a sprint before the day of a big race. Mental marathoners–such as yourself–should not study for any more than one hour 24 hours before a SSAT Upper Level test. That's because your brain requires some rest to be at its best. The night before your exam, spend some time with family or friends, or read a book.

- **Avoid bright screens** - You'll have to get some good shuteye the night before your test. Bright screens (such as the ones coming from your laptop, TV, or mobile device) should be avoided altogether. Staring at such a screen will keep your brain up, making it hard to drift asleep at a reasonable hour.

- **Make sure your dinner is healthy** - The meal that you have for dinner should be nutritious. Be sure to drink plenty of water as well. Load up on your complex carbohydrates, much like a marathon runner would do. Pasta, rice, and potatoes are ideal options here, as are vegetables and protein sources.

- **Get your bag ready for test day** - The night prior to your test, pack your bag with your stationery, admissions pass, ID, and any other gear that you need. Keep the bag right by your front door.

- **Make plans to reach the testing site** - Before going to sleep, ensure that you understand precisely how you will arrive at the site of the test. If parking is something you'll have to find first, plan for it. If you're dependent

on public transit, then review the schedule. You should also make sure that the train/bus/subway/streetcar you use will be running. Find out about road closures as well. If a parent or friend is accompanying you, ensure that they understand what steps they have to take as well.

The Day of the Test

- **Get up reasonably early, but not too early.**

- **Have breakfast** - Breakfast improves your concentration, memory, and mood. As such, make sure the breakfast that you eat in the morning is healthy. The last thing you want to be is distracted by a grumbling tummy. If it's not your own stomach making those noises, another test taker close to you might be instead. Prevent discomfort or embarrassment by consuming a healthy breakfast. Bring a snack with you if you think you'll need it.

- **Follow your daily routine** - Do you watch Good Morning America each morning while getting ready for the day? Don't break your usual habits on the day of the test. Likewise, if coffee isn't something you drink in the morning, then don't take up the habit hours before your test. Routine consistency lets you concentrate on the main objective–doing the best you can on your test.

- **Wear layers** - Dress yourself up in comfortable layers. You should be ready for any kind of internal temperature. If it gets too warm during the test, take a layer off.

- **Get there on time** - The last thing you want to do is get to the test site late. Rather, you should be there 45 minutes prior to the start of the test. Upon your arrival, try not to hang out with anybody who is nervous. Any anxious energy they exhibit shouldn't influence you.

- **Leave the books at home** - No books should be brought to the test site. If you start developing anxiety before the test, books could encourage you to do some last-minute studying, which will only hinder you. Keep the books far away–better yet, leave them at home.

- **Make your voice heard** - If something is off, speak to a proctor. If medical attention is needed or if you'll require anything, consult the proctor prior to the start of the test. Any doubts you have should be clarified. You should be entering the test site with a state of mind that is completely clear.

- **Have faith in yourself** - When you feel confident, you will be able to perform at your best. When you are waiting for the test to begin, envision yourself receiving an outstanding result. Try to see yourself as someone who knows all the answers, no matter what the questions are. A lot of athletes tend to use this technique–particularly before a big competition. Your expectations will be reflected by your performance.

During your test

- **Be calm and breathe deeply** - You need to relax before the test, and some deep breathing will go a long way to help you do that. Be confident and calm. You got this. Everybody feels a little stressed out just before an evaluation of any kind is set to begin. Learn some effective breathing exercises. Spend a minute meditating before the test starts. Filter out any negative thoughts you have. Exhibit confidence when having such thoughts.

- **Concentrate on the test** - Refrain from comparing yourself to anyone else. You shouldn't be distracted by the people near you or random noise. Concentrate exclusively on the test. If you find yourself irritated by surrounding noises, earplugs can be used to block sounds off close to you. Don't forget–the test is going to last several hours if you're taking more than one subject of the test. Some of that time will be dedicated to brief sections. Concentrate on the specific section you are working on during a particular moment. Do not let your mind wander off to upcoming or previous sections.

- **Skip challenging questions** - Optimize your time when taking the test. Lingering on a single question for too long will work against you. If you don't know what the answer is to a certain question, use your best guess, and mark the question so you can review it later on. There is no need to spend time attempting to solve something you aren't sure about. That time would be better served handling the questions you can actually answer well. You will not be penalized for getting the wrong answer on a test like this.

- **Try to answer each question individually** - Focus only on the question you are working on. Use one of the test-taking strategies to solve the problem. If you aren't able to come up with an answer, don't get frustrated. Simply skip that question, then move onto the next one.

- **Don't forget to breathe!** Whenever you notice your mind wandering, your stress levels boosting, or frustration brewing, take a thirty-second break. Shut your eyes, drop your pencil, breathe deeply, and let your shoulders relax. You will end up being more productive when you allow yourself to relax for a moment.

- **Review your answer.** If you still have time at the end of the test, don't waste it. Go back and check over your answers. It is worth going through the test from start to finish to ensure that you didn't make a sloppy mistake somewhere.

- **Optimize your breaks** - When break time comes, use the restroom, have a snack, and reactivate your energy for the subsequent section. Doing some stretches can help stimulate your blood flow.

After your test

- **Take it easy** - You will need to set some time aside to relax and decompress once the test has concluded. There is no need to stress yourself out about what you could've said, or what you may have done wrong. At this point, there's nothing you can do about it. Your energy and time would be better spent on something that will bring you happiness for the remainder of your day.

- **Redoing the test** - Did you pass the test? Congratulations! Your hard work paid off!

  If you have failed your test, though, don't worry! The test can be retaken. In such cases, you will need to follow the retake policy. You also need to re-register to take the exam again.

Contents

Chapter: Exponents and Variables   45

6

Multiplication Property of Exponents 46
Division Property of Exponents 47
Powers of Products and Quotients 48
Zero and Negative Exponents 49
Negative Exponents and Negative Bases 50
Scientific Notation 51
Radicals 52
Chapter 6: Practices 53
Chapter 6: Answers 56

Chapter: Expressions and Variables   57

7

Simplifying Variable Expressions 58
Simplifying Polynomial Expressions 59
The Distributive Property 60
Evaluating One Variable 61
Evaluating Two Variables 62
Chapter 7: Practices 63
Chapter 7: Answers 66

Chapter: Equations and Inequalities   67

8

One–Step Equations 68
Multi–Step Equations 69
System of Equations 70
Graphing Single–Variable Inequalities 71
One–Step Inequalities 72
Multi–Step Inequalities 73
Chapter 8: Practices 74
Chapter 8: Answers 76

Chapter: Lines and Slope   77

9

Finding Slope 78
Graphing Lines Using Slope–Intercept Form 79
Writing Linear Equations 80
Finding Midpoint 81
Finding Distance of Two Points 82
Graphing Linear Inequalities 83
Chapter 9: Practices 84
Chapter 9: Answers 86

CHAPTER

1 Fractions and Mixed Numbers

Math topics that you'll learn in this chapter:

- ☑ Simplifying Fractions
- ☑ Adding and Subtracting Fractions
- ☑ Multiplying and Dividing Fractions
- ☑ Adding Mixed Numbers
- ☑ Subtracting Mixed Numbers
- ☑ Multiplying Mixed Numbers
- ☑ Dividing Mixed Numbers

1

Simplifying Fractions

- A fraction contains two numbers separated by a bar between them. The bottom number, called the denominator, is the total number of equally divided portions in one whole. The top number, called the numerator, is how many portions you have. And the bar represents the operation of division.

- Simplifying a fraction means reducing it to the lowest terms. To simplify a fraction, evenly divide both the top and bottom of the fraction by $2, 3, 5, 7$, etc.

- Continue until you can't go any further.

Examples:

Example 1. Simplify $\frac{18}{30}$

Solution: To simplify $\frac{18}{30}$, find a number that both 18 and 30 are divisible by. Both are divisible by 6. Then: $\frac{18}{30} = \frac{18 \div 6}{30 \div 6} = \frac{3}{5}$

Example 2. Simplify $\frac{32}{80}$

Solution: To simplify $\frac{32}{80}$, find a number that both 32 and 80 are divisible by. Both are divisible by 8 and 16. Then: $\frac{32}{80} = \frac{32 \div 8}{80 \div 8} = \frac{4}{10}$, 4 and 10 are divisible by 2, then: $\frac{4}{10} = \frac{2}{5}$ or $\frac{32}{80} = \frac{32 \div 16}{80 \div 16} = \frac{2}{5}$

Example 3. Simplify $\frac{40}{120}$

Solution: To simplify $\frac{40}{120}$, find a number that both 40 and 120 are divisible by. Both are divisible by 40, then: $\frac{40}{120} = \frac{40 \div 40}{120 \div 40} = \frac{1}{3}$

Adding and Subtracting Fractions

- For "like" fractions (fractions with the same denominator), add or subtract the numerators (top numbers) and write the answer over the common denominator (bottom numbers).

- Adding and Subtracting fractions with the same denominator:

$$\frac{a}{b} + \frac{c}{b} = \frac{a+c}{b} \qquad\qquad \frac{a}{b} - \frac{c}{b} = \frac{a-c}{b}$$

- Find equivalent fractions with the same denominator before you can add or subtract fractions with different denominators.

- Adding and Subtracting fractions with different denominators:

$$\frac{a}{b} + \frac{c}{d} = \frac{ad+bc}{bd} \qquad\qquad \frac{a}{b} - \frac{c}{d} = \frac{ad-bc}{bd}$$

Examples:

Example 1. Find the sum. $\frac{2}{3} + \frac{1}{2} =$

Solution: These two fractions are "unlike" fractions. (they have different denominators). Use this formula: $\frac{a}{b} + \frac{c}{d} = \frac{ad+cb}{bd}$

Then: $\frac{2}{3} + \frac{1}{2} = \frac{(2)(2)+(3)(1)}{3 \times 2} = \frac{4+3}{6} = \frac{7}{6}$

Example 2. Find the difference. $\frac{3}{5} - \frac{2}{7} =$

Solution: For "unlike" fractions, find equivalent fractions with the same denominator before you can add or subtract fractions with different denominators. Use this formula: $\frac{a}{b} - \frac{c}{d} = \frac{ad-bc}{bd}$

$\frac{3}{5} - \frac{2}{7} = \frac{(3)(7)-(2)(5)}{5 \times 7} = \frac{21-10}{35} = \frac{11}{35}$

Multiplying and Dividing Fractions

- **Multiplying fractions:** multiply the top numbers and multiply the bottom numbers. Simplify if necessary. $\frac{a}{b} \times \frac{c}{d} = \frac{a \times c}{b \times d}$

- **Dividing fractions:** Keep, Change, Flip

- Keep the first fraction, change the division sign to multiplication, and flip the numerator and denominator of the second fraction. Then, solve!

$$\frac{a}{b} \div \frac{c}{d} = \frac{a}{b} \times \frac{d}{c} = \frac{a \times d}{b \times c}$$

Examples:

Example 1. Multiply. $\frac{2}{3} \times \frac{3}{5} =$

Solution: Multiply the top numbers and multiply the bottom numbers.
$\frac{2}{3} \times \frac{3}{5} = \frac{2 \times 3}{3 \times 5} = \frac{6}{15}$, now, simplify: $\frac{6}{15} = \frac{6 \div 3}{15 \div 3} = \frac{2}{5}$

Example 2. Solve. $\frac{3}{4} \div \frac{2}{5} =$

Solution: Keep the first fraction, change the division sign to multiplication, and flip the numerator and denominator of the second fraction.
Then: $\frac{3}{4} \div \frac{2}{5} = \frac{3}{4} \times \frac{5}{2} = \frac{3 \times 5}{4 \times 2} = \frac{15}{8}$

Example 3. Calculate. $\frac{4}{5} \times \frac{3}{4} =$

Solution: Multiply the top numbers and multiply the bottom numbers.
$\frac{4}{5} \times \frac{3}{4} = \frac{4 \times 3}{5 \times 4} = \frac{12}{20}$, simplify: $\frac{12}{20} = \frac{12 \div 4}{20 \div 4} = \frac{3}{5}$

Example 4. Solve. $\frac{5}{6} \div \frac{3}{7} =$

Solution: Keep the first fraction, change the division sign to multiplication, and flip the numerator and denominator of the second fraction.
Then: $\frac{5}{6} \div \frac{3}{7} = \frac{5}{6} \times \frac{7}{3} = \frac{5 \times 7}{6 \times 3} = \frac{35}{18}$

Adding Mixed Numbers

Use the following steps for adding mixed numbers:

- Add whole numbers of the mixed numbers.

- Add the fractions of the mixed numbers.

- Find the Least Common Denominator (LCD) if necessary.

- Add whole numbers and fractions.

- Write your answer in lowest terms.

Examples:

Example 1. Add mixed numbers. $2\frac{1}{2} + 1\frac{2}{3} =$

Solution: Let's rewriting our equation with parts separated, $2\frac{1}{2} + 1\frac{2}{3} = 2 + \frac{1}{2} + 1 + \frac{2}{3}$.
Now, add whole number parts: $2 + 1 = 3$
Add the fraction parts $\frac{1}{2} + \frac{2}{3}$. Rewrite to solve with the equivalent fractions. $\frac{1}{2} +$
$\frac{2}{3} = \frac{3}{6} + \frac{4}{6} = \frac{7}{6}$. The answer is an improper fraction (numerator is bigger than denominator). Convert the improper fraction into a mixed number: $\frac{7}{6} = 1\frac{1}{6}$. Now, combine the whole and fraction parts: $3 + 1\frac{1}{6} = 4\frac{1}{6}$

Example 2. Find the sum. $1\frac{3}{4} + 2\frac{1}{2} =$

Solution: Rewriting our equation with parts separated, $1 + \frac{3}{4} + 2 + \frac{1}{2}$. Add the whole number parts:
$1 + 2 = 3$. Add the fraction parts: $\frac{3}{4} + \frac{1}{2} = \frac{3}{4} + \frac{2}{4} = \frac{5}{4}$
Convert the improper fraction into a mixed number: $\frac{5}{4} = 1\frac{1}{4}$.
Now, combine the whole and fraction parts: $3 + 1\frac{1}{4} = 4\frac{1}{4}$

Subtracting Mixed Numbers

Use these steps for subtracting mixed numbers.

- Convert mixed numbers into improper fractions. $a\frac{c}{b} = \frac{ab+c}{b}$

- Find equivalent fractions with the same denominator for unlike fractions. (fractions with different denominators)

- Subtract the second fraction from the first one. $\frac{a}{b} - \frac{c}{d} = \frac{ad-bc}{bd}$

- Write your answer in lowest terms.

- If the answer is an improper fraction, convert it into a mixed number.

Examples:

Example 1. Subtract. $2\frac{1}{3} - 1\frac{1}{2} =$

Solution: Convert mixed numbers into fractions: $2\frac{1}{3} = \frac{2\times3+1}{3} = \frac{7}{3}$ and $1\frac{1}{2} = \frac{1\times2+1}{2} = \frac{3}{2}$

These two fractions are "unlike" fractions. (they have different denominators). Find equivalent fractions with the same denominator. Use this formula: $\frac{a}{b} - \frac{c}{d} = \frac{ad-bc}{bd}$

$\frac{7}{3} - \frac{3}{2} = \frac{(7)(2)-(3)(3)}{3\times2} = \frac{14-9}{6} = \frac{5}{6}$

Example 2. Find the difference. $3\frac{4}{7} - 2\frac{3}{4} =$

Solution: Convert mixed numbers into fractions: $3\frac{4}{7} = \frac{3\times7+4}{7} = \frac{25}{7}$ and $2\frac{3}{4} = \frac{2\times4+3}{4} = \frac{11}{4}$

Then: $3\frac{4}{7} - 2\frac{3}{4} = \frac{25}{7} - \frac{11}{4} = \frac{(25)(4)-(11)(7)}{7\times4} = \frac{23}{28}$

Multiplying Mixed Numbers

Use the following steps for multiplying mixed numbers:

- Convert the mixed numbers into fractions. $a\frac{c}{b} = a + \frac{c}{b} = \frac{ab+c}{b}$

- Multiply fractions. $\frac{a}{b} \times \frac{c}{d} = \frac{a \times c}{b \times d}$

- Write your answer in lowest terms.

- If the answer is an improper fraction (numerator is bigger than denominator), convert it into a mixed number.

Examples:

Example 1. Multiply. $4\frac{1}{2} \times 2\frac{2}{5} =$

Solution: Convert mixed numbers into fractions, $4\frac{1}{2} = \frac{4 \times 2 + 1}{2} = \frac{9}{2}$ and $2\frac{2}{5} = \frac{2 \times 5 + 2}{5} =$ $\frac{12}{5}$. Apply the fractions rule for multiplication: $\frac{9}{2} \times \frac{12}{5} = \frac{9 \times 12}{2 \times 5} = \frac{108}{10} = \frac{54}{5}$ The answer is an improper fraction. Convert it into a mixed number. $\frac{54}{5} = 10\frac{4}{5}$

Example 2. Multiply. $3\frac{2}{3} \times 2\frac{5}{6} =$

Solution: Converting mixed numbers into fractions, $3\frac{2}{3} \times 2\frac{5}{6} = \frac{11}{3} \times \frac{17}{6}$ Apply the fractions rule for multiplication: $\frac{11}{3} \times \frac{17}{6} = \frac{11 \times 17}{3 \times 6} = \frac{187}{18} = 10\frac{7}{18}$

Example 3. Find the product. $5\frac{1}{4} \times 3\frac{3}{8} =$

Solution: Convert mixed numbers to fractions: $5\frac{1}{4} = \frac{21}{4}$ and $3\frac{3}{8} = \frac{27}{8}$. Multiply two fractions:

$\frac{21}{4} \times \frac{27}{8} = \frac{21 \times 27}{4 \times 8} = \frac{567}{32} = 17\frac{23}{32}$

Dividing Mixed Numbers

Use the following steps for dividing mixed numbers:

- Convert the mixed numbers into fractions. $a\frac{c}{b} = a + \frac{c}{b} = \frac{ab+c}{b}$

- Divide fractions: Keep, Change, Flip: Keep the first fraction, change the division sign to multiplication, and flip the numerator and denominator of the second fraction. Then, solve! $\frac{a}{b} \div \frac{c}{d} = \frac{a}{b} \times \frac{d}{c} = \frac{a \times d}{b \times c}$

- Write your answer in lowest terms.

- If the answer is an improper fraction (numerator is bigger than denominator), convert it into a mixed number.

Examples:

Example 1. Solve. $2\frac{1}{3} \div 1\frac{1}{2}$

Solution: Convert mixed numbers into fractions: $2\frac{1}{3} = \frac{2\times3+1}{3} = \frac{7}{3}$ and $1\frac{1}{2} = \frac{1\times2+1}{2} = \frac{3}{2}$
Keep, Change, Flip: $\frac{7}{3} \div \frac{3}{2} = \frac{7}{3} \times \frac{2}{3} = \frac{7\times2}{3\times3} = \frac{14}{9}$. The answer is an improper fraction.
Convert it into a mixed number: $\frac{14}{9} = 1\frac{5}{9}$

Example 2. Solve. $3\frac{3}{4} \div 2\frac{2}{5}$

Solution: Convert mixed numbers to fractions, then solve:
$3\frac{3}{4} \div 2\frac{2}{5} = \frac{15}{4} \div \frac{12}{5} = \frac{15}{4} \times \frac{5}{12} = \frac{75}{48} = 1\frac{9}{16}$

Example 3. Solve. $2\frac{4}{5} \div 1\frac{2}{3}$

Solution: Converting mixed numbers to fractions: $2\frac{4}{5} \div 1\frac{2}{3} = \frac{14}{5} \div \frac{5}{3}$
Keep, Change, Flip: $\frac{14}{5} \div \frac{5}{3} = \frac{14}{5} \times \frac{3}{5} = \frac{14\times3}{5\times5} = \frac{42}{25} = 1\frac{17}{25}$

Chapter 1: Practices

✎ Simplify each fraction.

1) $\frac{2}{8} =$

2) $\frac{5}{15} =$

3) $\frac{10}{90} =$

4) $\frac{12}{16} =$

5) $\frac{25}{45} =$

6) $\frac{42}{54} =$

7) $\frac{48}{60} =$

8) $\frac{52}{169} =$

✎ Find the sum or difference.

9) $\frac{3}{10} + \frac{2}{10} =$

10) $\frac{4}{9} - \frac{1}{9} =$

11) $\frac{2}{3} + \frac{6}{15} =$

12) $\frac{17}{24} - \frac{5}{8} =$

13) $\frac{7}{54} - \frac{1}{9} =$

14) $\frac{4}{5} - \frac{1}{6} =$

15) $\frac{6}{7} - \frac{3}{8} =$

16) $\frac{2}{13} + \frac{1}{4} =$

✎ Find the products or quotients.

17) $\frac{2}{9} \div \frac{4}{3} =$

18) $\frac{14}{5} \div \frac{28}{35} =$

19) $\frac{9}{25} \times \frac{5}{27} =$

20) $\frac{65}{72} \times \frac{12}{15} =$

✎ Find the sum.

21) $2\frac{1}{5} + 1\frac{2}{5} =$

22) $5\frac{1}{9} + 2\frac{7}{9} =$

23) $2\frac{3}{4} + 1\frac{1}{8} =$

24) $2\frac{2}{7} + 4\frac{1}{21} =$

25) $5\frac{3}{5} + 1\frac{4}{9} =$

26) $3\frac{3}{11} + 4\frac{6}{7} =$

✎ Find the difference.

27) $5\frac{1}{3} - 4\frac{2}{3} =$ 31) $4\frac{3}{4} - 2\frac{1}{28} =$

28) $4\frac{7}{10} - 1\frac{3}{10} =$ 32) $4\frac{2}{7} - 3\frac{1}{6} =$

29) $3\frac{1}{3} - 2\frac{2}{9} =$ 33) $5\frac{3}{10} - 3\frac{3}{4} =$

30) $6\frac{1}{2} - 3\frac{1}{3} =$ 34) $6\frac{9}{20} - 2\frac{1}{3} =$

✎ Find the products.

35) $1\frac{1}{2} \times 2\frac{3}{7} =$ 39) $2\frac{1}{5} \times 5\frac{1}{2} =$

36) $1\frac{3}{4} \times 1\frac{3}{5} =$ 40) $2\frac{1}{2} \times 4\frac{4}{5} =$

37) $4\frac{1}{2} \times 1\frac{5}{6} =$ 41) $3\frac{1}{5} \times 4\frac{1}{2} =$

38) $1\frac{2}{7} \times 3\frac{1}{5} =$ 42) $4\frac{9}{10} \times 4\frac{1}{2} =$

✎ Solve.

43) $1\frac{1}{3} \div 1\frac{2}{3} =$ 47) $4\frac{1}{5} \div 2\frac{2}{3} =$

44) $2\frac{1}{4} \div 1\frac{1}{2} =$ 48) $1\frac{2}{3} \div 1\frac{3}{8} =$

45) $5\frac{1}{3} \div 3\frac{1}{2} =$ 49) $4\frac{1}{2} \div 2\frac{2}{3} =$

46) $3\frac{2}{7} \div 1\frac{1}{8} =$ 50) $1\frac{2}{11} \div 1\frac{1}{8} =$

Chapter 1: Answers

1) $\frac{1}{4}$

2) $\frac{1}{3}$

3) $\frac{1}{9}$

4) $\frac{3}{4}$

5) $\frac{5}{9}$

6) $\frac{7}{9}$

7) $\frac{4}{5}$

8) $\frac{4}{13}$

9) $\frac{1}{2}$

10) $\frac{1}{3}$

11) $\frac{16}{15} = 1\frac{1}{15}$

12) $\frac{1}{12}$

13) $\frac{1}{54}$

14) $\frac{19}{30}$

15) $\frac{27}{56}$

16) $\frac{21}{52}$

17) $\frac{1}{6}$

18) $\frac{7}{2} = 3\frac{1}{2}$

19) $\frac{1}{15}$

20) $\frac{13}{18}$

21) $3\frac{3}{5}$

22) $7\frac{8}{9}$

23) $3\frac{7}{8}$

24) $6\frac{1}{3}$

25) $7\frac{2}{45}$

26) $8\frac{10}{77}$

27) $\frac{2}{3}$

28) $3\frac{2}{5}$

29) $1\frac{1}{9}$

30) $3\frac{1}{6}$

31) $2\frac{5}{7}$

32) $1\frac{5}{42}$

33) $1\frac{11}{20}$

34) $4\frac{7}{60}$

35) $3\frac{9}{14}$

36) $2\frac{4}{5}$

37) $8\frac{1}{4}$

38) $4\frac{4}{35}$

39) $12\frac{1}{10}$

40) 12

41) $14\frac{2}{5}$

42) $22\frac{1}{20}$

43) $\frac{4}{5}$

44) $1\frac{1}{2}$

45) $1\frac{11}{21}$

46) $2\frac{58}{63}$

47) $1\frac{23}{40}$

48) $1\frac{7}{33}$

49) $1\frac{11}{16}$

50) $1\frac{5}{99}$

CHAPTER

2 Decimals

Math topics that you'll learn in this chapter:

- ☑ Comparing Decimals
- ☑ Rounding Decimals
- ☑ Adding and Subtracting Decimals
- ☑ Multiplying and Dividing Decimals

13

Comparing Decimals

- A decimal is a fraction written in a special form. For example, instead of writing $\frac{1}{2}$ you can write: 0.5

- A Decimal Number contains a Decimal Point. It separates the whole number part from the fractional part of a decimal number.

- Let's review decimal place values: Example: **45.3861**

4: tens	5: ones	3: tenths
8: hundredths	6: thousandths	1: tens thousandths

- To compare two decimals, compare each digit of two decimals in the same place value. Start from left. Compare hundreds, tens, ones, tenth, hundredth, etc.

- To compare numbers, use these symbols:

Equal to $=$	Less than $<$	Greater than $>$
Greater than or equal $\geq$	Less than or equal $\leq$	

Examples:

Example 1. Compare 0.03 and 0.30.

Solution: 0.30 *is greater than* 0.03, because the tenth place of 0.30 is 3, but the tenth place of 0.03 is zero. Then: $0.03 < 0.30$

Example 2. Compare 0.0917 and 0.217.

Solution: 0.217 *is greater than* 0.0917, because the tenth place of 0.217 is 2, but the tenth place of 0.0917 is zero. Then: $0.0917 < 0.217$

Rounding Decimals

- We can round decimals to a certain accuracy or number of decimal places. This is used to make calculations easier to do and results easier to understand when exact values are not too important.

- First, you'll need to remember your place values: For example: **12.4869**

  1: tens   2: ones   4: tenths

  8: hundredths   6: thousandths   9: tens thousandths

- To round a decimal, first find the place value you'll round to.

- Find the digit to the right of the place value you're rounding to. If it is 5 or bigger, add 1 to the place value you're rounding to and remove all digits on its right side. If the digit to the right of the place value is less than 5, keep the place value and remove all digits on the right.

Examples:

Example 1. Round 4.3679 to the thousandth place value.

Solution: First, look at the next place value to the right, (tens thousandths). It's 9 and it is greater than 5. Thus add 1 to the digit in the thousandth place. The thousandth place is 7. $\rightarrow 7 + 1 = 8$, then, the answer is 4.368

Example 2. Round 1.5237 to the nearest hundredth.

Solution: First, look at the digit to the right of hundredth (thousandths place value). It's 3 and it is less than 5, thus remove all the digits to the right of hundredth place. Then, the answer is 1.52

Adding and Subtracting Decimals

- Line up the decimal numbers.

- Add zeros to have the same number of digits for both numbers if necessary.

- Remember your place values: For example: 73.5196

  7: tens   3: ones   5: tenths

  1: hundredths   9: thousandths   6: tens thousandths

- Add or subtract using column addition or subtraction.

Examples:

Example 1. Add. 1.7 + 4.12

Solution: First, line up the numbers: $\begin{array}{r} 1.7 \\ + 4.12 \\ \hline \end{array}$ → Add a zero to have the same number of digits for both numbers. $\begin{array}{r} 1.70 \\ + 4.12 \\ \hline \end{array}$ → Start with the hundredths place: 0 + 2 = 2, $\begin{array}{r} 1.70 \\ + 4.12 \\ \hline 2 \end{array}$ → Continue with tenths place: 7 + 1 = 8, $\begin{array}{r} 1.70 \\ + 4.12 \\ \hline .82 \end{array}$ → Add the ones place: 4 + 1 = 5, $\begin{array}{r} 1.70 \\ + 4.12 \\ \hline 5.82 \end{array}$ The answer is 5.82.

Example 2. Find the difference. 5.58 − 4.23

Solution: First, line up the numbers: $\begin{array}{r} 5.58 \\ - 4.23 \\ \hline \end{array}$ → Start with the hundredths place: 8 − 3 = 5, $\begin{array}{r} 5.58 \\ - 4.23 \\ \hline 5 \end{array}$ → Continue with tenths place. 5 − 2 = 3, $\begin{array}{r} 5.58 \\ - 4.23 \\ \hline .35 \end{array}$ → Subtract the ones place. 5 − 4 = 1, $\begin{array}{r} 5.58 \\ - 4.23 \\ \hline 1.35 \end{array}$

Multiplying and Dividing Decimals

For multiplying decimals:

- Ignore the decimal point and set up and multiply the numbers as you do with whole numbers.

- Count the total number of decimal places in both of the factors.

- Place the decimal point in the product.

For dividing decimals:

- If the divisor is not a whole number, move the decimal point to the right to make it a whole number. Do the same for the dividend.

- Divide similar to whole numbers.

Examples:

Example 1. Find the product. $0.65 \times 0.24 =$

Solution: Set up and multiply the numbers as you do with whole numbers. Line up the numbers: $\begin{array}{r} 65 \\ \times\,24 \\ \hline \end{array} \rightarrow$ Start with the ones place then continue with other digits $\rightarrow \dfrac{\begin{array}{r} 65 \\ \times 24 \end{array}}{1{,}560}$. Count the total number of decimal places in both of the factors. There are four decimals digits. (two for each factor 0.65 and 0.24) Then: $0.65 \times 0.24 = 0.1560 = 0.156$

Example 2. Find the quotient. $1.20 \div 0.4 =$

Solution: The divisor is not a whole number. Multiply it by 10 to get 4: $\rightarrow 0.4 \times 10 = 4$

Do the same for the dividend to get 12. $\rightarrow 1.20 \times 10 = 12$

Now, divide $12 \div 4 = 3$. The answer is 3.

Chapter 2: Practices

✍ **Compare. Use >, =, and <**

1) 0.5 ☐ 0.6

2) 0.9 ☐ 0.8

3) 0.1 ☐ 0.2

4) 0.02 ☐ 0.06

5) 0.05 ☐ 0.08

6) 0.12 ☐ 0.09

7) 3.2 ☐ 2.5

8) 4.8 ☐ 8.4

9) 0.005 ☐ 0.05

10) 2.02 ☐ 20.020

11) 55.100 ☐ 55.10

12) 0.44 ☐ 0.440

13) 6.01 ☐ 6.0100

14) 0.77 ☐ 77.0

✍ **Round each decimal to the nearest whole number.**

15) 5.8

16) 6.4

17) 12.3

18) 9.2

19) 7.6

20) 22.4

21) 6.8

22) 15.9

23) 13.41

24) 16.78

25) 67.58

26) 42.67

27) 55.89

28) 14.32

29) 78.88

30) 98.29

✍ Find the sum or difference.

31) $12.1 + 36.2 =$

32) $56.3 - 22.2 =$

33) $45.1 + 12.8 =$

34) $27.9 - 16.4 =$

35) $98.8 - 56.6 =$

36) $28.45 + 13.22 =$

37) $16.78 + 45.11 =$

38) $86.16 - 72.12 =$

39) $96.23 - 28.32 =$

40) $57.33 + 67.46 =$

41) $46.26 - 39.49 =$

42) $44.95 + 76.53 =$

43) $79.37 - 52.89 =$

44) $19.99 + 28.7 =$

45) $83.48 - 49.3 =$

46) $19.6 + 42.98 =$

✍ Find the product or quotient.

47) $3.3 \times 0.2 =$

48) $2.4 \div 0.3 =$

49) $8.1 \times 1.4 =$

50) $4.8 \div 0.2 =$

51) $4.1 \times 0.3 =$

52) $8.6 \div 0.2 =$

53) $9.9 \times 0.8 =$

54) $1.84 \div 0.2 =$

55) $2.1 \times 8.4 =$

56) $1.6 \times 4.5 =$

57) $9.2 \times 3.1 =$

58) $36.6 \div 1.6 =$

59) $1.91 \times 5.2 =$

60) $3.65 \times 1.4 =$

61) $24.82 \div 0.4 =$

62) $12.4 \times 4.20 =$

Chapter 2: Answers

1) <	22) 16	43) 26.48
2) >	23) 13	44) 48.69
3) <	24) 17	45) 34.18
4) <	25) 68	46) 62.58
5) <	26) 43	47) 0.66
6) >	27) 56	48) 8
7) >	28) 14	49) 11.34
8) <	29) 79	50) 24
9) <	30) 98	51) 1.23
10) <	31) 48.3	52) 43
11) =	32) 34.1	53) 7.92
12) =	33) 57.9	54) 9.2
13) =	34) 11.5	55) 17.64
14) <	35) 42.2	56) 7.2
15) 6	36) 41.67	57) 28.52
16) 6	37) 61.89	58) 22.875
17) 12	38) 14.04	59) 9.932
18) 9	39) 67.91	60) 5.11
19) 8	40) 124.79	61) 62.05
20) 22	41) 6.77	62) 52.08
21) 7	42) 121.48	

3 Integers and Order of Operations

Math topics that you'll learn in this chapter:

☑ Adding and Subtracting Integers

☑ Multiplying and Dividing Integers

☑ Order of Operations

☑ Integers and Absolute Value

21

Adding and Subtracting Integers

- Integers include zero, counting numbers, and the negative of the counting numbers. $\{... ,-3,-2,-1,0,1,2,3, ...\}$

- Add a positive integer by moving to the right on the number line. (you will get a bigger number)

- Add a negative integer by moving to the left on the number line. (you will get a smaller number)

- Subtract an integer by adding its opposite.

Number line

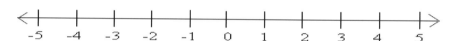

Examples:

Example 1. Solve. $(-2)-(-8) =$

Solution: Keep the first number and convert the sign of the second number to its opposite. (change subtraction into addition. Then: $(-2) + 8 = 6$

Example 2. Solve. $4 + (5 - 10) =$

Solution: First, subtract the numbers in brackets, $5 - 10 = -5$.
Then: $4 + (-5) = \rightarrow$ change addition into subtraction: $4 - 5 = -1$

Example 3. Solve. $(9-14) + 15 =$

Solution: First, subtract the numbers in brackets, $9 - 14 = -5$
Then: $-5 + 15 = \rightarrow -5 + 15 = 10$

Example 4. Solve. $12 + (-3 - 10) =$

Solution: First, subtract the numbers in brackets, $-3 - 10 = -13$
Then: $12 + (-13) = \rightarrow$ change addition into subtraction: $12 - 13 = -1$

Multiplying and Dividing Integers

Use the following rules for multiplying and dividing integers:

-   (negative) × (negative) = positive

-   (negative) ÷ (negative) = positive

-   (negative) × (positive) = negative

-   (negative) ÷ (positive) = negative

-   (positive) × (positive) = positive

-   (positive) ÷ (negative) = negative

Examples:

Example 1. Solve. $3 \times (-4) =$

Solution: Use this rule: (positive) × (negative) = negative.
Then: $(3) \times (-4) = -12$

Example 2. Solve. $(-3) + (-24 \div 3) =$

Solution: First, divide -24 by 3, the numbers in brackets, use this rule:
(negative) ÷ (positive) = negative. Then: $-24 \div 3 = -8$
$(-3) + (-24 \div 3) = (-3) + (-8) = -3 - 8 = -11$

Example 3. Solve. $(12 - 15) \times (-2) =$

Solution: First, subtract the numbers in brackets,
$12 - 15 = -3 \rightarrow (-3) \times (-2) =$
Now use this rule: (negative) × (negative) = positive $\rightarrow (-3) \times (-2) = 6$

Example 4. Solve. $(12 - 8) \div (-4) =$

Solution: First, subtract the numbers in brackets,
$12 - 8 = 4 \rightarrow (4) \div (-4) =$
Now use this rule: (positive) ÷ (negative) = negative $\rightarrow (4) \div (-4) = -1$

Order of Operations

- In Mathematics, "operations" are addition, subtraction, multiplication, division, exponentiation (written as b^n), and grouping.

- When there is more than one math operation in an expression, use PEMDAS: (to memorize this rule, remember the phrase "Please Excuse My Dear Aunt Sally".)

❖ Parentheses

❖ Exponents

❖ Multiplication and Division (from left to right)

❖ Addition and Subtraction (from left to right)

Examples:

Example 1. Calculate. $(2 + 6) \div (2^2 \div 4) =$

Solution: First, simplify inside parentheses:
$(8) \div (4 \div 4) = (8) \div (1)$, Then: $(8) \div (1) = 8$

Example 2. Solve. $(6 \times 5) - (14 - 5) =$

Solution: First, calculate within parentheses: $(6 \times 5) - (14 - 5) = (30) - (9)$, Then: $(30) - (9) = 21$

Example 3. Calculate. $-4[(3 \times 6) \div (9 \times 2)] =$

Solution: First, calculate within parentheses:
$-4[(18) \div (9 \times 2)] = -4[(18) \div (18)] = -4[1]$
multiply -4 and 1. Then: $-4[1] = -4$

Example 4. Solve. $(28 \div 7) + (-19 + 3) =$

Solution: First, calculate within parentheses:
$(28 \div 7) + (-19 + 3) = (4) + (-16)$ Then: $(4) - (16) = -12$

Integers and Absolute Value

- The absolute value of a number is its distance from zero, in either direction, on the number line. For example, the distance of 9 and -9 from zero on number line is 9.

- The absolute value of an integer is the numerical value without its sign. (negative or positive)

- The vertical bar is used for absolute value as in $|x|$.

- The absolute value of a number is never negative; because it only shows, "how far the number is from zero".

Examples:

Example 1. Calculate. $|14 - 2| \times 5 =$

Solution: First, solve $|14 - 2|$, $\rightarrow |14 - 2| = |12|$, the absolute value of 12 is 12, $|12| = 12$, Then: $12 \times 5 = 60$

Example 2. Solve. $\frac{|-24|}{4} \times |5 - 7| =$

Solution: First, find $|-24| \rightarrow$ the absolute value of -24 is 24. Then: $|-24| = 24$, $\frac{24}{4} \times |5 - 7| =$

Now, calculate $|5 - 7|$, $\rightarrow |5 - 7| = |-2|$, the absolute value of -2 is 2. $|-2| = 2$ Then: $\frac{24}{4} \times 2 = 6 \times 2 = 12$

Example 3. Solve. $|8 - 2| \times \frac{|-4 \times 7|}{2} =$

Solution: First, calculate $|8 - 2|$, $\rightarrow |8 - 2| = |6|$, the absolute value of 6 is 6, $|6| = 6$. Then: $6 \times \frac{|-4 \times 7|}{2}$

Now calculate $|-4 \times 7|$, $\rightarrow |-4 \times 7| = |-28|$, the absolute value of -28 is 28, $|-28| = 28$, Then: $6 \times \frac{28}{2} = 6 \times 14 = 84$

Chapter 3: Practices

✍ Find each sum or difference.

1) $-9 + 16 =$

2) $-18 - 6 =$

3) $-24 + 10 =$

4) $30 + (-5) =$

5) $15 + (-3) =$

6) $(-13) + (-4) =$

7) $25 + (3 - 10) =$

8) $12 - (-6 + 9) =$

9) $5 - (-2 + 7) =$

10) $(-11) + (-5 + 6) =$

11) $(-3) + (9 - 16) =$

12) $(-8) - (13 + 4) =$

13) $(-7 + 9) - 39 =$

14) $(-30 + 6) - 14 =$

15) $(-5 + 9) + (-3 + 7) =$

16) $(8 - 19) - (-4 + 12) =$

17) $(-9 + 2) - (6 - 7) =$

18) $(-12 - 5) - (-4 - 14) =$

✍ Solve.

19) $3 \times (-6) =$

20) $(-32) \div 4 =$

21) $(-5) \times 4 =$

22) $(25) \div (-5) =$

23) $(-72) \div 8 =$

24) $(-2) \times (-6) \times 5 =$

25) $(-2) \times 3 \times (-7) =$

26) $(-1) \times (-3) \times (-5) =$

27) $(-2) \times (-3) \times (-6) =$

28) $(-12 + 3) \times (-5) =$

29) $(-3 + 4) \times (-11) =$

30) $(-9) \times (6 - 5) =$

31) $(-3 - 7) \times (-6) =$

32) $(-7 + 3) \times (-9 + 6) =$

33) $(-15) \div (-17 + 12) =$

34) $(-3 - 2) \times (-9 + 7) =$

35) $(-15 + 31) \div (-2) =$

36) $(-64) \div (-16 + 8) =$

✍ Evaluate each expression.

37) $3 + (2 \times 5) =$

38) $(5 \times 4) - 7 =$

39) $(-9 \times 2) + 6 =$

40) $(7 \times 3) - (-5) =$

41) $(-8) + (2 \times 7) =$

42) $(9 - 6) + (3 \times 4) =$

43) $(-19 + 5) + (6 \times 2) =$

44) $(32 \div 4) + (1 - 13) =$

45) $(-36 \div 6) - (12 + 3) =$

46) $(-16 + 5) - (54 \div 9) =$

47) $(-20 + 4) - (35 \div 5) =$

48) $(42 \div 7) + (2 \times 3) =$

49) $(28 \div 4) + (2 \times 6) =$

50) $2[(3 \times 3) - (4 \times 5)] =$

51) $3[(2 \times 8) + (4 \times 3)] =$

52) $2[(9 \times 3) - (6 \times 4)] =$

53) $4[(4 \times 8) \div (4 \times 4)] =$

54) $-5[(10 \times 8) \div (5 \times 8)] =$

✍ Find the answers.

55) $|-5| + |7 - 10| =$

56) $|-4 + 6| + |-2| =$

57) $|-9| + |1 - 9| =$

58) $|-7| - |8 - 12| =$

59) $|9 - 11| + |8 - 15| =$

60) $|-7 + 10| - |-8 + 3| =$

61) $|-12 + 6| - |3 - 9| =$

62) $5 + |2 - 6| + |3 - 4| =$

63) $-4 + |2 - 6| + |1 - 9| =$

64) $|-6| \times |-7| + |2 - 8| =$

65) $|-12| \times |-3| + |4 - 28| =$

66) $|4 \times (-2)| \times |-9| =$

67) $|-3 \times 2| \times |-5| =$

68) $|3 - 12| - |-3 \times 7| =$

69) $|-9| + |-7 \times 5| =$

70) $|-11| + |-6 \times 4| =$

71) $|-4 \times 2 + 6| \times |-2 \times 8| =$

72) $|-1 \times 5 + 2| \times |-4| =$

Chapter 3: Answers

1) 7	25) 42	49) 19
2) −24	26) −15	50) −22
3) −14	27) −36	51) 84
4) 25	28) 45	52) 6
5) 12	29) −11	53) 8
6) −17	30) −9	54) −10
7) 18	31) 60	55) 8
8) 9	32) 12	56) 4
9) 0	33) 3	57) 17
10) −10	34) 10	58) 3
11) −10	35) −8	59) 9
12) −25	36) 8	60) −2
13) −37	37) 13	61) 0
14) −38	38) 13	62) 10
15) 8	39) −12	63) 8
16) −19	40) 26	64) 48
17) −6	41) 6	65) 60
18) 1	42) 15	66) 72
19) −18	43) −2	67) 30
20) −8	44) −4	68) −12
21) −20	45) −21	69) 44
22) −5	46) −17	70) 35
23) −9	47) −23	71) 32
24) 60	48) 12	72) 12

4 Ratios and Proportions

Math topics that you'll learn in this chapter:

- ☑ Simplifying Ratios
- ☑ Proportional Ratios
- ☑ Similarity and Ratios

29

Simplifying Ratios

- Ratios are used to make comparisons between two numbers.

- Ratios can be written as a fraction, using the word "to", or with a colon. Example: $\frac{3}{4}$ or "3 to 4" or 3:4

- You can calculate equivalent ratios by multiplying or dividing both sides of the ratio by the same number.

Examples:

Example 1. Simplify. $8:2 =$

Solution: Both numbers 8 and 2 are divisible by $2 \Rightarrow 8 \div 2 = 4$, $4 \div 2 = 2$, Then: $8:2 = 4:1$

Example 2. Simplify. $\frac{9}{33} =$

Solution: Both numbers 9 and 33 are divisible by $3 \Rightarrow 33 \div 3 = 11$, $9 \div 3 = 3$, Then: $\frac{9}{33} = \frac{3}{11}$

Example 3. There are 24 students in a class and 10 are girls. Find the ratio of girls to boys in that class.

Solution: Subtract 10 from 24 to find the number of boys in the class. $24 - 10 = 14$. There are 14 boys in the class. So, the ratio of girls to boys is 10:14. Now, simplify this ratio. Both 14 and 10 are divisible by 2. Then: $14 \div 2 = 7$, and $10 \div 2 = 5$. In the simplest form, this ratio is 5:7

Example 4. A recipe calls for butter and sugar in the ratio 3:4. If you're using 9 cups of butter, how many cups of sugar should you use?

Solution: Since you use 9 cups of butter, or 3 times as much, you need to multiply the amount of sugar by 3. Then: $4 \times 3 = 12$. So, you need to use 12 cups of sugar. You can solve this using equivalent fractions: $\frac{3}{4} = \frac{9}{12}$

Proportional Ratios

- Two ratios are proportional if they represent the same relationship.

- A proportion means that two ratios are equal. It can be written in two ways: $\dfrac{a}{b} = \dfrac{c}{d}$ $a : b = c : d$

- The proportion $\dfrac{a}{b} = \dfrac{c}{d}$ can be written as: $a \times d = c \times b$

Examples:

Example 1. Solve this proportion for x. $\dfrac{2}{5} = \dfrac{6}{x}$

Solution: Use cross multiplication: $\dfrac{2}{5} = \dfrac{6}{x} \Rightarrow 2 \times x = 6 \times 5 \Rightarrow 2x = 30$

Divide both sides by 2 to find x: $x = \dfrac{30}{2} \Rightarrow x = 15$

Example 2. If a box contains red and blue balls in ratio of $3 : 5$ red to blue, how many red balls are there if 45 blue balls are in the box?

Solution: Write a proportion and solve. $\dfrac{3}{5} = \dfrac{x}{45}$

Use cross multiplication: $3 \times 45 = 5 \times x \Rightarrow 135 = 5x$

Divide to find x: $x = \dfrac{135}{5} \Rightarrow x = 27$. There are 27 red balls in the box.

Example 3. Solve this proportion for x. $\dfrac{4}{9} = \dfrac{16}{x}$

Solution: Use cross multiplication: $\dfrac{4}{9} = \dfrac{16}{x} \Rightarrow 4 \times x = 9 \times 16 \Rightarrow 4x = 144$

Divide to find x: $x = \dfrac{144}{4} \Rightarrow x = 36$

Example 4. Solve this proportion for x. $\dfrac{5}{7} = \dfrac{20}{x}$

Solution: Use cross multiplication: $\dfrac{5}{7} = \dfrac{20}{x} \Rightarrow 5 \times x = 7 \times 20 \Rightarrow 5x = 140$

Divide to find x: $x = \dfrac{140}{5} \Rightarrow x = 28$

Similarity and Ratios

- Two figures are similar if they have the same shape.

- Two or more figures are similar if the corresponding angles are equal, and the corresponding sides are in proportion.

Examples:

Example 1. The following triangles are similar. What is the value of the unknown side?

Example 2.

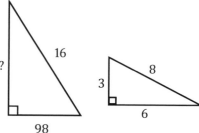

Solution: Find the corresponding sides and write a proportion.
$\frac{8}{16} = \frac{6}{x}$. Now, use the cross product to solve for x:
$\frac{8}{16} = \frac{6}{x} \rightarrow 8 \times x = 16 \times 6 \rightarrow 8x = 96$. Divide both sides
by 8. Then: $8x = 96 \rightarrow x = \frac{96}{8} \rightarrow x = 12$

The missing side is 12.

Example 3. Two rectangles are similar. The first is 5 feet wide and 15 feet long. The second is 10 feet wide. What is the length of the second rectangle?

Solution: Let's put x for the length of the second rectangle. Since two rectangles are similar, their corresponding sides are in proportion. Write a proportion and solve for the missing number.
$\frac{5}{10} = \frac{15}{x} \rightarrow 5x = 10 \times 15 \rightarrow 5x = 150 \rightarrow x = \frac{150}{5} = 30$
The length of the second rectangle is 30 feet.

Chapter 4: Practices

✎ Reduce each ratio.

1) $2 : 18 =$ ___ : ___

2) $5 : 35 =$ ___ : ___

3) $8 : 72 =$ ___ : ___

4) $24 : 36 =$ ___ : ___

5) $25 : 40 =$ ___ : ___

6) $40 : 72 =$ ___ : ___

7) $28 : 63 =$ ___ : ___

8) $18 : 81 =$ ___ : ___

9) $13 : 52 =$ ___ : ___

10) $56 : 72 =$ ___ : ___

11) $42 : 63 =$ ___ : ___

12) $32 : 96 =$ ___ : ___

✎ Solve.

13) Bob has 16 red cards and 20 green cards. What is the ratio of Bob's red cards to his green cards? _____

14) In a party, 34 soft drinks are required for every 20 guests. If there are 260 guests, how many soft drinks are required? _____

15) Sara has 56 blue pens and 28 black pens. What is the ratio of Sara's black pens to her blue pens? _____

16) In Jack's class, 48 of the students are tall and 20 are short. In Michael's class 28 students are tall and 12 students are short. Which class has a higher ratio of tall to short students? _____

17) The price of 6 apples at the Quick Market is $1.52. The price of 5 of the same apples at Walmart is $1.32. Which place is the better buy? _____

18) The bakers at a Bakery can make 180 bagels in 6 hours. How many bagels can they bake in 16 hours? What is that rate per hour? _____

19) You can buy 6 cans of green beans at a supermarket for $3.48. How much does it cost to buy 38 cans of green beans? _____

✒ Solve each proportion.

20) $\frac{3}{2} = \frac{9}{x} \Rightarrow x = $ ____

21) $\frac{7}{2} = \frac{x}{4} \Rightarrow x = $ ____

22) $\frac{1}{3} = \frac{2}{x} \Rightarrow x = $ ____

23) $\frac{1}{4} = \frac{5}{x} \Rightarrow x = $ ____

24) $\frac{9}{6} = \frac{x}{2} \Rightarrow x = $ ____

25) $\frac{3}{6} = \frac{5}{x} \Rightarrow x = $ ____

26) $\frac{7}{x} = \frac{2}{6} \Rightarrow x = $ ____

27) $\frac{2}{x} = \frac{4}{10} \Rightarrow x = $ ____

28) $\frac{3}{2} = \frac{x}{8} \Rightarrow x = $ ____

29) $\frac{x}{6} = \frac{5}{3} \Rightarrow x = $ ____

30) $\frac{3}{9} = \frac{5}{x} \Rightarrow x = $ ____

31) $\frac{4}{18} = \frac{2}{x} \Rightarrow x = $ ____

32) $\frac{6}{16} = \frac{3}{x} \Rightarrow x = $ ____

33) $\frac{2}{5} = \frac{x}{20} \Rightarrow x = $ ____

34) $\frac{28}{8} = \frac{x}{2} \Rightarrow x = $ ____

35) $\frac{3}{5} = \frac{x}{15} \Rightarrow x = $ ____

36) $\frac{2}{7} = \frac{x}{14} \Rightarrow x = $ ____

37) $\frac{x}{18} = \frac{3}{2} \Rightarrow x = $ ____

38) $\frac{x}{24} = \frac{2}{6} \Rightarrow x = $ ____

39) $\frac{5}{x} = \frac{4}{20} \Rightarrow x = $ ____

40) $\frac{10}{x} = \frac{20}{80} \Rightarrow x = $ ____

41) $\frac{90}{6} = \frac{x}{2} \Rightarrow x = $ ____

✒ Solve each problem.

42) Two rectangles are similar. The first is 8 *feet* wide and 32 *feet* long. The second is 12 *feet* wide. What is the length of the second rectangle?

_____

43) Two rectangles are similar. One is 4.6 *meters* by 7 *meters*. The longer side of the second rectangle is 28 *meters*. What is the other side of the second rectangle? _____

Chapter 4: Answers

1) $1:9$

2) $1:7$

3) $1:9$

4) $2:3$

5) $5:8$

6) $5:9$

7) $4:9$

8) $2:9$

9) $1:4$

10) $7:9$

11) $2:3$

12) $1:3$

13) $4:5$

14) 442

15) $1:2$

16) *Jack's class*: $\frac{48}{20} = \frac{12}{5}$ *Michael's class*: $\frac{28}{12} = \frac{7}{3}$ Jack's class has a higher ratio of tall to short student: $\frac{12}{5} > \frac{7}{3}$

17) Quick market

18) 480, 30 bagels per hour

19) $22.04

20) 6

21) 14

22) 6

23) 20

24) 3

25) 10

26) 21

27) 5

28) 12

29) 10

30) 15

31) 9

32) 8

33) 8

34) 7

35) 9

36) 4

37) 27

38) 8

39) 25

40) 40

41) 30

42) 48 *meters*

43) 18.4 *meters*

5 Percentage

Math topics that you'll learn in this chapter:

- ☑ Percent Problems
- ☑ Percent of Increase and Decrease
- ☑ Discount, Tax and Tip
- ☑ Simple Interest

37

Percent Problems

- Percent is a ratio of a number and 100. It always has the same denominator, 100. The percent symbol is "%".

- Percent means "per 100". So, 20% is $\frac{20}{100}$.

- In each percent problem, we are looking for the base, or the part or the percent.

- Use these equations to find each missing section in a percent problem:

  ❖ Base = Part ÷ Percent
  ❖ Part = Percent × Base
  ❖ Percent = Part ÷ Base

Examples:

Example 1. What is 20% of 40?

Solution: In this problem, we have the percent (20%) and the base (40) and we are looking for the "part". Use this formula: *Part = Percent × Base*.
Then: $Part = 20\% \times 40 = \frac{20}{100} \times 40 = 0.20 \times 40 = 8$. The answer: 20% of 40 is 8.

Example 2. 25 is what percent of 500?

Solution: In this problem, we are looking for the percent. Use this equation:
$Percent = Part \div Base \rightarrow Percent = 25 \div 500 = 0.05 = 5\%$.
Then: 25 is 5 percent of 500.

Example 3. 80 is 20 percent of what number?

Solution: In this problem, we are looking for the base. Use this equation:
$Base = Part \div Percent \rightarrow Base = 80 \div 20\% = 80 \div 0.20 = 400$
Then: 80 is 20 percent of 400.

Percent of Increase and Decrease

- Percent of change (increase or decrease) is a mathematical concept that represents the degree of change over time.

- To find the percentage of increase or decrease:

  1. New Number – Original Number

  2. (The result ÷ Original Number) × 100

- Or use this formula: Percent of change $= \frac{new\ number - original\ number}{original\ number} \times 100$

- Note: If your answer is a negative number, then this is a percentage decrease. If it is positive, then this is a percentage increase.

Examples:

Example 1. The price of a shirt increases from \$30 to \$36. What is the percentage increase?

Solution: First, find the difference: $36 - 30 = 6$

Then: $(6 \div 30) \times 100 = \frac{6}{30} \times 100 = 20$. The percentage increase is 20%. It means that the price of the shirt increased by 20%.

Example 2. The price of a table decreased from \$50 to \$35. What is the percent of decrease?

Solution: Use this formula:

$Percent\ of\ change = \dfrac{new\ number - original\ number}{original\ number} \times 100 =$

$\frac{35-50}{50} \times 100 = \frac{-15}{50} \times 100 = -30$. The percentage decrease is 30. (the negative sign means percentage decrease) Therefore, the price of the table decreased by 30%.

Discount, Tax and Tip

- To find the discount: Multiply the regular price by the rate of discount

- To find the selling price: Original price – discount

- To find tax: Multiply the tax rate to the taxable amount (income, property value, etc.)

- To find the tip, multiply the rate to the selling price.

Examples:

Example 1. With an 20% discount, Ella saved $50 on a dress. What was the original price of the dress?

Solution: let x be the original price of the dress. Then: 20 % *of* $x = 50$. Write an equation and solve for x: $0.20 \times x = 50 \to x = \frac{50}{0.20} = 250$. The original price of the dress was $250.

Example 2. Sophia purchased a new computer for a price of $820 at the Apple Store. What is the total amount her credit card is charged if the sales tax is 5%?

Solution: The taxable amount is $820, and the tax rate is 5%. Then:
$$Tax = 0.05 \times 820 = 41$$
Final price = Selling price + Tax → final price = $820 + $41 = $861

Example 3. Nicole and her friends went out to eat at a restaurant. If their bill was $60.00 and they gave their server a 15% tip, how much did they pay altogether?

Solution: First, find the tip. To find the tip, multiply the rate to the bill amount. $Tip = 60 \times 0.15 = 9$. The final price is: $60 + $9 = $69

Simple Interest

- Simple Interest: The charge for borrowing money or the return for lending it.

- Simple interest is calculated on the initial amount (principal).

- To solve a simple interest problem, use this formula:

$$Interest = principal \times rate \times time \quad (I = p \times r \times t = prt)$$

Examples:

Example 1. Find simple interest for $200 investment at 5% for 3 years.

Solution: Use Interest formula:
$I = prt$ ($P = \$200$, $r = 5\% = \frac{5}{100} = 0.05$ and $t = 3$)
Then: $I = 200 \times 0.05 \times 3 = \$30$

Example 2. Find simple interest for $1,200 at 8% for 6 years.

Solution: Use Interest formula:
$I = prt$ ($P = \$1,200$, $r = 8\% = \frac{8}{100} = 0.08$ and $t = 6$)
Then: $I = 1,200 \times 0.08 \times 6 = \$576$

Example 3. Andy received a student loan to pay for his educational expenses this year. What is the interest on the loan if he borrowed $4,500 at 6% for 5 years?

Solution: Use Interest formula: $I = prt$. $P = \$4,500$, $r = 6\% = 0.06$ and $t = 5$
Then: $I = 4,500 \times 0.06 \times 5 = \$1,350$

Example 4. Bob is starting his own small business. He borrowed $20,000 from the bank at an 8% rate for 6 months. Find the interest Bob will pay on this loan.

Solution: Use Interest formula:
$I = prt$. $P = \$20,000$, $r = 8\% = 0.08$ and $t = 0.5$ (6 months is half year). Then:
$$I = 20,000 \times 0.08 \times 0.5 = \$800$$

Chapter 5: Practices

✑ Solve each problem.

1) What is 15% of 60? ____

2) What is 55% of 800? ____

3) What is 22% of 120? ____

4) What is 18% of 40? ____

5) 90 is what percent of 200? ____%

6) 30 is what percent of 150? ____%

7) 14 is what percent of 250? ____%

8) 60 is what percent of 300? ____%

9) 30 is 120 percent of what number? ____

10) 120 is 20 percent of what number? ____

11) 15 is 5 percent of what number? ____

12) 22 is 20% of what number? ____

✑ Solve each problem.

13) Bob got a raise, and his hourly wage increased from $15 to $21. What is the percent increase? _____ %

14) The price of a pair of shoes increases from $32 to $36. What is the percent increase? ___ %

15) At a coffeeshop, the price of a cup of coffee increased from $1.35 to $1.62. What is the percent increase in the cost of the coffee? _____ %

16) A $45 shirt now selling for $36 is discounted by what percent? _____ %

17) Joe scored 30 out of 35 marks in Algebra, 20 out of 30 marks in science and 58 out of 70 marks in mathematics. In which subject his percentage of marks is best? _____

18) Emma purchased a computer for $420. The computer is regularly priced at $480. What was the percent discount Emma received on the computer? _____

19) A chemical solution contains 15% alcohol. If there is 54 ml of alcohol, what is the volume of the solution? _____

✎ Find the selling price of each item.

20) Original price of a computer: $600

    Tax: 8%, Selling price: $_____

21) Original price of a laptop: $450

    Tax: 10%, Selling price: $_____

22) Nicolas hired a moving company. The company charged $500 for its services, and Nicolas gives the movers a 14% tip. How much does Nicolas tip the movers? $_____

23) Mason has lunch at a restaurant and the cost of his meal is $40. Mason wants to leave a 20% tip. What is Mason's total bill, including tip? $_____

✎ Determine the simple interest for the following loans.

24) $1,000 *at* 5% *for* 4 *years.* $___

25) $400 *at* 3% *for* 5 *years.* $___

26) $240 *at* 4% *for* 3 *years.* $___

27) $500 at 4.5% for 6 years. $___

✎ Solve.

28) A new car, valued at $20,000, depreciates at 8% per year. What is the value of the car one year after purchase? $_____

29) Sara puts $7,000 into an investment yielding 3% annual simple interest; she left the money in for five years. How much interest does Sara get at the end of those five years? $_____

Chapter 5: Answers

1) 9

2) 440

3) 26.4

4) 7.2

5) 45%

6) 20%

7) 5.6%

8) 20%

9) 25

10) 600

11) 300

12) 110

13) 40%

14) 12.5%

15) 20%

16) 20%

17) Algebra

18) 12.5%

19) 360 ml

20) $648.00

21) $495.00

22) $70.00

23) $48.00

24) $200.00

25) $60.00

26) $28.80

27) $135.00

28) $18.400

29) $1,050

CHAPTER

6 | Exponents and Variables

Math topics that you'll learn in this chapter:

☑ Multiplication Property of Exponents

☑ Division Property of Exponents

☑ Powers of Products and Quotients

☑ Zero and Negative Exponents

☑ Negative Exponents and Negative Bases

☑ Scientific Notation

☑ Radicals

45

Multiplication Property of Exponents

- Exponents are shorthand for repeated multiplication of the same number by itself. For example, instead of 2×2, we can write 2^2. For $3 \times 3 \times 3 \times 3$, we can write 3^4

- In algebra, a variable is a letter used to stand for a number. The most common letters are: $x, y, z, a, b, c, m,$ and n.

- Exponent's rules: $x^a \times x^b = x^{a+b}$, $\frac{x^a}{x^b} = x^{a-b}$

$$(x^a)^b = x^{a \times b} \qquad\qquad (xy)^a = x^a \times y^a \qquad\qquad \left(\frac{a}{b}\right)^c = \frac{a^c}{b^c}$$

Examples:

Example 1. Multiply. $2x^2 \times 3x^4$

Solution: Use Exponent's rules: $x^a \times x^b = x^{a+b} \rightarrow x^2 \times x^4 = x^{2+4} = x^6$
Then: $2x^2 \times 3x^4 = 6x^6$

Example 2. Simplify. $(x^4y^2)^2$

Solution: Use Exponent's rules: $(x^a)^b = x^{a \times b}$.
Then: $(x^4y^2)^2 = x^{4 \times 2}y^{2 \times 2} = x^8y^4$

Example 3. Multiply. $5x^8 \times 6x^5$

Solution: Use Exponent's rules: $x^a \times x^b = x^{a+b} \rightarrow x^8 \times x^5 = x^{8+5} = x^{13}$
Then: $5x^8 \times 6x^5 = 30x^{13}$

Example 4. Simplify. $(x^4y^7)^3$

Solution: Use Exponent's rules: $(x^a)^b = x^{a \times b}$.
Then: $(x^4y^7)^3 = x^{4 \times 3}y^{7 \times 3} = x^{12}y^{21}$

Division Property of Exponents

For division of exponents use following formulas:

- $\frac{x^a}{x^b} = x^{a-b} \ (x \neq 0)$

- $\frac{x^a}{x^b} = \frac{1}{x^{b-a}} , \ (x \neq 0)$

- $\frac{1}{x^b} = x^{-b}$

Examples:

Example 1. Simplify. $\frac{16x^3y}{2xy^2} =$

Solution: First, cancel the common factor: $2 \rightarrow \frac{16x^3y}{2xy^2} = \frac{8x^3y}{xy^2}$

Use Exponent's rules: $\frac{x^a}{x^b} = x^{a-b} \rightarrow \frac{x^3}{x} = x^{3-1} = x^2$ and $\frac{x^a}{x^b} = \frac{1}{x^{b-a}} \rightarrow \frac{y}{y^2} = \frac{1}{y^{2-1}} = \frac{1}{y}$

Then: $\frac{16x^3y}{2xy^2} = \frac{8x^2}{y}$

Example 2. Simplify. $\frac{24x^8}{3x^6} =$

Solution: Use Exponent's rules: $\frac{x^a}{x^b} = x^{a-b} \rightarrow \frac{x^8}{x^6} = x^{8-6} = x^2$

Then: $\frac{24x^8}{3x^6} = 8x^2$

Example 3. Simplify. $\frac{7x^4y^2}{28x^3y} =$

Solution: First, cancel the common factor: $7 \rightarrow \frac{x^4y^2}{4x^3y}$

Use Exponent's rules: $\frac{x^a}{x^b} = x^{a-b} \rightarrow \frac{x^4}{x^3} = x^{4-3} = x$ and $\frac{y^2}{y} = y$

Then: $\frac{7x^4y^2}{28x^3y} = \frac{xy}{4}$

Powers of Products and Quotients

- For any nonzero numbers a and b and any integer x, $(ab)^x = a^x \times b^x$
  and $\left(\frac{a}{b}\right)^c = \frac{a^c}{b^c}$

Examples:

Example 1. Simplify. $(3x^3y^2)^2$

Solution: Use Exponent's rules: $(x^a)^b = x^{a \times b}$

$(3x^3y^2)^2 = (3)^2(x^3)^2(y^2)^2 = 9x^{3 \times 2}y^{2 \times 2} = 9x^6y^4$

Example 2. Simplify. $\left(\frac{2x^3}{3x^2}\right)^2$

Solution: First, cancel the common factor: $x \to \left(\frac{2x^3}{3x^2}\right) = \left(\frac{2x}{3}\right)^2$

Use Exponent's rules: $\left(\frac{a}{b}\right)^c = \frac{a^c}{b^c}$, Then: $\left(\frac{2x}{3}\right)^2 = \frac{(2x)^2}{(3)^2} = \frac{4x^2}{9}$

Example 3. Simplify. $\left(-4x^3y^5\right)^2$

Solution: Use Exponent's rules: $(x^a)^b = x^{a \times b}$

$$\left(-4x^3y^5\right)^2 = (-4)^2(x^3)^2\left(y^5\right)^2 = 16x^{3 \times 2}y^{5 \times 2} = 16x^6y^{10}$$

Example 4. Simplify. $\left(\frac{5x}{4x^2}\right)^2$

Solution: First, cancel the common factor: $x \to \left(\frac{5x}{4x^2}\right)^2 = \left(\frac{5}{4x}\right)^2$

Use Exponent's rules: $\left(\frac{a}{b}\right)^c = \frac{a^c}{b^c}$, Then: $\left(\frac{5}{4x}\right)^2 = \frac{5^2}{(4x)^2} = \frac{25}{16x^2}$

Zero and Negative Exponents

- Zero-Exponent Rule: $a^0 = 1$, this means that anything raised to the zero power is 1. For example: $(5xy)^0 = 1$ (number zero is an exception: $0^0 = 0$)

- A negative exponent simply means that the base is on the wrong side of the fraction line, so you need to flip the base to the other side. For instance, "x^{-2}" (pronounced as "ecks to the minus two") just means "x^2" but underneath, as in $\frac{1}{x^2}$.

Examples:

Example 1. Evaluate. $\left(\frac{4}{5}\right)^{-2} =$

Solution: Use negative exponent's rule: $\left(\frac{x^a}{x^b}\right)^{-2} = \left(\frac{x^b}{x^a}\right)^2 \rightarrow \left(\frac{4}{5}\right)^{-2} = \left(\frac{5}{4}\right)^2$

Then: $\left(\frac{5}{4}\right)^2 = \frac{5^2}{4^2} = \frac{25}{16}$

Example 2. Evaluate. $\left(\frac{3}{2}\right)^{-3} =$

Solution: Use negative exponent's rule: $\left(\frac{x^a}{x^b}\right)^{-3} = \left(\frac{x^b}{x^a}\right)^3 \rightarrow \left(\frac{3}{2}\right)^{-3} = \left(\frac{2}{3}\right)^3 =$

Then: $\left(\frac{2}{3}\right)^3 = \frac{2^3}{3^3} = \frac{8}{27}$

Example 3. Evaluate. $\left(\frac{a}{b}\right)^0 =$

Solution: Use zero-exponent Rule: $a^0 = 1$

Then: $\left(\frac{a}{b}\right)^0 = 1$

Example 4. Evaluate. $\left(\frac{4}{7}\right)^{-1} =$

Solution: Use negative exponent's rule: $\left(\frac{x^a}{x^b}\right)^{-1} = \left(\frac{x^b}{x^a}\right)^1 \rightarrow \left(\frac{4}{7}\right)^{-1} = \left(\frac{7}{4}\right)^1 = \frac{7}{4}$

Negative Exponents and Negative Bases

- A negative exponent is the reciprocal of that number with a positive exponent. $(3)^{-2} = \frac{1}{3^2}$

- To simplify a negative exponent, make the power positive!

- The parenthesis is important! -5^{-2} is not the same as $(-5)^{-2}$

$$-5^{-2} = -\frac{1}{5^2} \text{ and } (-5)^{-2} = +\frac{1}{5^2}$$

Examples:

Example 1. Simplify. $\left(\frac{2a}{3c}\right)^{-2} =$

Solution: Use negative exponent's rule: $\left(\frac{x^a}{x^b}\right)^{-2} = \left(\frac{x^b}{x^a}\right)^{2} \rightarrow \left(\frac{2a}{3c}\right)^{-2} = \left(\frac{3c}{2a}\right)^{2}$

Now use exponent's rule: $\left(\frac{a}{b}\right)^{c} = \frac{a^c}{b^c} \rightarrow = \left(\frac{3c}{2a}\right)^{2} = \frac{3^2 c^2}{2^2 a^2}$

Then: $\frac{3^2 c^2}{2^2 a^2} = \frac{9c^2}{4a^2}$

Example 2. Simplify. $\left(\frac{x}{4y}\right)^{-3} =$

Solution: Use negative exponent's rule: $\left(\frac{x^a}{x^b}\right)^{-3} = \left(\frac{x^b}{x^a}\right)^{3} \rightarrow \left(\frac{x}{4y}\right)^{-3} = \left(\frac{4y}{x}\right)^{3}$

Now use exponent's rule: $\left(\frac{a}{b}\right)^{c} = \frac{a^c}{b^c} \rightarrow \left(\frac{4y}{x}\right)^{3} = \frac{4^3 y^3}{x^3} = \frac{64y^3}{x^3}$

Example 3. Simplify. $\left(\frac{5a}{2c}\right)^{-2} =$

Solution: Use negative exponent's rule: $\left(\frac{x^a}{x^b}\right)^{-2} = \left(\frac{x^b}{x^a}\right)^{2} \rightarrow \left(\frac{5a}{2c}\right)^{-2} = \left(\frac{2c}{5a}\right)^{2}$

Now use exponent's rule: $\left(\frac{a}{b}\right)^{c} = \frac{a^c}{b^c} \rightarrow = \left(\frac{2c}{5a}\right)^{2} = \frac{2^2 c^2}{5^2 a^2}$

Then: $\frac{2^2 c^2}{5^2 a^2} = \frac{4c^2}{25a^2}$

Scientific Notation

- Scientific notation is used to write very big or very small numbers in decimal form.

- In scientific notation, all numbers are written in the form of: $m \times 10^n$, where m is greater than 1 and less than 10.

- To convert a number from scientific notation to standard form, move the decimal point to the left (if the exponent of ten is a negative number), or to the right (if the exponent is positive).

Examples:

Example 1. Write 0.00024 in scientific notation.

Solution: First, move the decimal point to the right so you have a number between 1 and 10. That number is 2.4. Now, determine how many places the decimal moved in step 1 by the power of 10. We moved the decimal point 4 digits to the right. Then: $10^{-4} \rightarrow$ When the decimal moved to the right, the exponent is negative. Then: $0.00024 = 2.4 \times 10^{-4}$

Example 2. Write 3.8×10^{-5} in standard notation.

Solution: The exponent is negative 5. Then, move the decimal point to the left five digits. (remember $3.8 = 0000003.8$) When the decimal moved to the right, the exponent is negative. Then: $3.8 \times 10^{-5} = 0.000038$

Example 3. Write 0.00031 in scientific notation.

Solution: First, move the decimal point to the right so you have a number between 1 and 10. Then: $m = 3.1$, Now, determine how many places the decimal moved in step 1 by the power of 10. $10^{-4} \rightarrow$ Then: $0.00031 = 3.1 \times 10^{-4}$

Example 4. Write 6.2×10^5 in standard notation.

Solution: $10^5 \rightarrow$ The exponent is positive 5. Then, move the decimal point to the right five digits. (remember $6.2 = 6.20000$) Then: $6.2 \times 10^5 = 620,000$

Radicals

- If n is a positive integer and x is a real number, then: $\sqrt[n]{x} = x^{\frac{1}{n}}$,

$$\sqrt[n]{xy} = x^{\frac{1}{n}} \times y^{\frac{1}{n}}, \ \sqrt[n]{\frac{x}{y}} = \frac{x^{\frac{1}{n}}}{y^{\frac{1}{n}}}, \text{ and } \sqrt[n]{x} \times \sqrt[n]{y} = \sqrt[n]{xy}$$

- A square root of x is a number r whose square is: $r^2 = x$ (r is a square root of x)

- To add and subtract radicals, we need to have the same values under the radical. For example: $\sqrt{3} + \sqrt{3} = 2\sqrt{3}$, $3\sqrt{5} - \sqrt{5} = 2\sqrt{5}$

Examples:

Example 1. Find the square root of $\sqrt{121}$.

Solution: First, factor the number: $121 = 11^2$, Then: $\sqrt{121} = \sqrt{11^2}$,
Now use radical rule: $\sqrt[n]{a^n} = a$. Then: $\sqrt{121} = \sqrt{11^2} = 11$

Example 2. Evaluate. $\sqrt{4} \times \sqrt{16} =$

Solution: Find the values of $\sqrt{4}$ and $\sqrt{16}$. Then: $\sqrt{4} \times \sqrt{16} = 2 \times 4 = 8$

Example 3. Solve. $5\sqrt{2} + 9\sqrt{2}$.

Solution: Since we have the same values under the radical, we can add these two radicals: $5\sqrt{2} + 9\sqrt{2} = 14\sqrt{2}$

Example 4. Evaluate. $\sqrt{2} \times \sqrt{50} =$

Solution: Use this radical rule: $\sqrt[n]{x} \times \sqrt[n]{y} = \sqrt[n]{xy} \rightarrow \sqrt{2} \times \sqrt{50} = \sqrt{100}$
The square root of 100 is 10. Then: $\sqrt{2} \times \sqrt{50} = \sqrt{100} = 10$

Chapter 6: Practices

✎ Find the products.

1) $x^2 \times 4xy^2 =$

2) $3x^2y \times 5x^3y^2 =$

3) $6x^4y^2 \times x^2y^3 =$

4) $7xy^3 \times 2x^2y =$

5) $-5x^5y^5 \times x^3y^2 =$

6) $-8x^3y^2 \times 3x^3y^2 =$

7) $-6x^2y^6 \times 5x^4y^2 =$

8) $-3x^3y^3 \times 2x^3y^2 =$

9) $-6x^5y^3 \times 4x^4y^3 =$

10) $-2x^4y^3 \times 5x^6y^2 =$

11) $-7y^6 \times 3x^6y^3 =$

12) $-9x^4 \times 2x^4y^2 =$

✎ Simplify.

13) $\frac{5^3 \times 5^4}{5^9 \times 5} =$

14) $\frac{3^3 \times 3^2}{7^2 \times 7} =$

15) $\frac{15x^5}{5x^3} =$

16) $\frac{16x^3}{4x^5} =$

17) $\frac{72y^2}{8x^3y^6} =$

18) $\frac{10x^3y^4}{50x^2y^3} =$

19) $\frac{13y^2}{52x^4y^4} =$

20) $\frac{50xy^3}{200x^3y^4} =$

21) $\frac{48x^2}{56x^2y^2} =$

22) $\frac{81y^6x}{54x^4y^3} =$

✎ Solve.

23) $(x^3y^3)^2 =$

24) $(3x^3y^4)^3 =$

25) $(4x \times 6xy^3)^2 =$

26) $(5x \times 2y^3)^3 =$

27) $\left(\frac{9x}{x^3}\right)^2 =$

28) $\left(\frac{3y}{18y^2}\right)^2 =$

29) $\left(\frac{3x^2y^3}{24x^4y^2}\right)^3 =$

30) $\left(\frac{26x^5y^3}{52x^3y^5}\right)^2 =$

31) $\left(\frac{18x^7y^4}{72x^5y^2}\right)^2 =$

32) $\left(\frac{12x^6y^4}{48x^5y^3}\right)^2 =$

✎ Evaluate each expression. (Zero and Negative Exponents)

33) $\left(\frac{1}{4}\right)^{-2} =$

34) $\left(\frac{1}{3}\right)^{-2} =$

35) $\left(\frac{1}{7}\right)^{-3} =$

36) $\left(\frac{2}{5}\right)^{-3} =$

37) $\left(\frac{2}{3}\right)^{-3} =$

38) $\left(\frac{3}{5}\right)^{-4} =$

✎ Write each expression with positive exponents.

39) $x^{-7} =$

40) $3y^{-5} =$

41) $15y^{-3} =$

42) $-20x^{-4} =$

43) $12a^{-3}b^5 =$

44) $25a^3b^{-4}c^{-3} =$

45) $-4x^5y^{-3}z^{-6} =$

46) $\frac{18y}{x^3y^{-2}} =$

47) $\frac{20a^{-2}b}{-12c^{-4}}$

✎ Write each number in scientific notation.

48) $0.00412 =$

49) $0.000053 =$

50) $66,000 =$

51) $72,000,000 =$

✎ Evaluate.

52) $\sqrt{8} \times \sqrt{8} =$ ⎯⎯⎯⎯

53) $\sqrt{36} - \sqrt{9} =$ ⎯⎯⎯

54) $\sqrt{81} + \sqrt{16} =$ ⎯⎯⎯

55) $\sqrt{4} \times \sqrt{25} =$ ⎯⎯⎯

56) $\sqrt{2} \times \sqrt{32} =$ ⎯⎯⎯

57) $4\sqrt{3} + 5\sqrt{3} =$ ⎯⎯⎯

Chapter 6: Answers

1) $4x^3y^2$

2) $15x^5y^3$

3) $6x^6y^5$

4) $14x^3y^4$

5) $-5x^8y^7$

6) $-24x^6y^4$

7) $-30x^6y^8$

8) $-6x^6y^5$

9) $-24x^9y^6$

10) $-10x^{10}y^5$

11) $-21x^6y^9$

12) $-18x^8y^2$

13) $\frac{1}{125}$

14) $\frac{243}{343}$

15) $3x^2$

16) $\frac{4}{x^2}$

17) $\frac{9}{x^3y^4}$

18) $\frac{xy}{5}$

19) $\frac{1}{4x^4y^2}$

20) $\frac{1}{4x^2y}$

21) $\frac{6}{7y^2}$

22) $\frac{3y^3}{2x^3}$

23) x^6y^6

24) $27x^9y^{12}$

25) $576x^4y^6$

26) $1,000x^3y^9$

27) $\frac{81}{x^4}$

28) $\frac{1}{36y^2}$

29) $\frac{y^3}{512x^6}$

30) $\frac{x^4}{4y^4}$

31) $\frac{x^4y^4}{16}$

32) $\frac{x^2y^2}{16}$

33) 16

34) 9

35) 343

36) $\frac{125}{8}$

37) $\frac{27}{8}$

38) $\frac{625}{81}$

39) $\frac{1}{x^7}$

40) $\frac{3}{y^5}$

41) $\frac{15}{y^3}$

42) $-\frac{20}{x^4}$

43) $\frac{12b^5}{a^3}$

44) $\frac{25a^3}{b^4c^3}$

45) $-\frac{4x^5}{y^3z^6}$

46) $\frac{18y^3}{x^3}$

47) $-\frac{5bc^4}{3a^2}$

48) 4.12×10^{-3}

49) 5.3×10^{-5}

50) 6.6×10^4

51) 7.2×10^7

52) 8

53) 3

54) 13

55) 10

56) 8

57) $9\sqrt{3}$

CHAPTER

7 Expressions and Variables

Math topics that you'll learn in this chapter:

☑ Simplifying Variable Expressions

☑ Simplifying Polynomial Expressions

☑ The Distributive Property

☑ Evaluating One Variable

☑ Evaluating Two Variables

57

Simplifying Variable Expressions

- In algebra, a variable is a letter used to stand for a number. The most common letters are $x, y, z, a, b, c, m,$ and n.

- An algebraic expression is an expression that contains integers, variables, and math operations such as addition, subtraction, multiplication, division, etc.

- In an expression, we can combine "like" terms. (values with same variable and same power)

Examples:

Example 1. Simplify. $(4x + 2x + 4) =$

Solution: In this expression, there are three terms: $4x$, $2x$, and 4. Two terms are "like terms": $4x$ and $2x$. Combine like terms. $4x + 2x = 6x$. Then: $(4x + 2x + 4) = 6x + 4$ (*remember you cannot combine variables and numbers.*)

Example 2. Simplify. $-2x^2 - 5x + 4x^2 - 9 =$

Solution: Combine "like" terms: $-2x^2 + 4x^2 = 2x^2$.
Then: $-2x^2 - 5x + 4x^2 - 9 = 2x^2 - 5x - 9$.

Example 3. Simplify. $(-8 + 6x^2 + 3x^2 + 9x) =$

Solution: Combine like terms. Then:
$(-8 + 6x^2 + 3x^2 + 9x) = 9x^2 + 9x - 8$

Example 4. Simplify. $-10x + 6x^2 - 3x + 9x^2 =$

Solution: Combine "like" terms: $-10x - 3x = -13x$, and $6x^2 + 9x^2 = 15x^2$
Then: $-10x + 6x^2 - 3x + 9x^2 = -13x + 15x^2$. Write in standard form (biggest powers first): $-13x + 15x^2 = 15x^2 - 13x$

Simplifying Polynomial Expressions

- In mathematics, a polynomial is an expression consisting of variables and coefficients that involves only the operations of addition, subtraction, multiplication, and non–negative integer exponents of variables.

$$P(x) = a_n x^n + a_{n-1} x^{n-1} + \ldots + a_2 x^2 + a_1 x + a_0$$

- Polynomials must always be simplified as much as possible. It means you must add together any like terms. (values with same variable and same power)

Examples:

Example 1. Simplify this Polynomial Expressions. $3x^2 - 6x^3 - 2x^3 + 4x^4$

Solution: Combine "like" terms: $-6x^3 - 2x^3 = -8x^3$
Then: $3x^2 - 6x^3 - 2x^3 + 4x^4 = 3x^2 - 8x^3 + 4x^4$
Now, write the expression in standard form: $3x^2 - 8x^3 + 4x^4 = 4x^4 - 8x^3 + 3x^2$

Example 2. Simplify this expression. $(-5x^2 + 2x^3) - (3x^3 - 6x^2) =$

Solution: First, multiply $(-)$ into $(3x^3 - 6x^2)$:
$(-5x^2 + 2x^3) - (3x^3 - 6x^2) = -5x^2 + 2x^3 - 3x^3 + 6x^2$
Then combine "like" terms: $-5x^2 + 2x^3 - 3x^3 + 6x^2 = x^2 - x^3$
And write in standard form: $x^2 - x^3 = -x^3 + x^2$

Example 3. Simplify. $3x^3 - 9x^4 - 8x^2 + 12x^4 =$

Solution: Combine "like" terms: $-9x^4 + 12x^4 = 3x^4$
Then: $3x^3 - 9x^4 - 8x^2 + 12x^4 = 3x^3 + 3x^4 - 8x^2$
And write in standard form: $3x^3 + 3x^4 - 8x^2 = 3x^4 + 3x^3 - 8x^2$

The Distributive Property

- The distributive property (or the distributive property of multiplication over addition and subtraction) simplifies and solves expressions in the form of: $a(b + c)$ or $a(b - c)$

- The distributive property is multiplying a term outside the parentheses by the terms inside.

- Distributive Property rule: $a(b + c) = ab + ac$

Examples:

Example 1. Simply using the distributive property. $(-2)(x + 3)$

Solution: Use Distributive Property rule: $a(b + c) = ab + ac$
$(-2)(x + 3) = (-2 \times x) + (-2) \times (3) = -2x - 6$

Example 2. Simply. $(-5)(-2x - 6)$

Solution: Use Distributive Property rule: $a(b + c) = ab + ac$
$(-5)(-2x - 6) = (-5 \times -2x) + (-5) \times (-6) = 10x + 30$

Example 3. Simply. $(7)(2x - 8) - 12x$

Solution: First, simplify $(7)(2x - 8)$ using the distributive property.
Then: $(7)(2x - 8) = 14x - 56$
Now combine like terms: $(7)(2x - 8) - 12x = 14x - 56 - 12x$
In this expression, $14x$ and $-12x$ are "like terms" and we can combine them.
$14x - 12x = 2x$. Then: $14x - 56 - 12x = 2x - 56$

Evaluating One Variable

- To evaluate one variable expressions, find the variable and substitute a number for that variable.

- Perform the arithmetic operations.

Examples:

Example 1. Calculate this expression for $x = 2$. $8 + 2x$

Solution: First, substitute 2 for x.

Then: $8 + 2x = 8 + 2(2)$

Now, use order of operation to find the answer: $8 + 2(2) = 8 + 4 = 12$

Example 2. Evaluate this expression for $x = -1$. $4x - 8$

Solution: First, substitute -1 for x.

Then: $4x - 8 = 4(-1) - 8$

Now, use order of operation to find the answer: $4(-1) - 8 = -4 - 8 = -12$

Example 3. Find the value of this expression when $x = 4$. $(16 - 5x)$

Solution: First, substitute 4 for x,

Then: $16 - 5x = 16 - 5(4) = 16 - 20 = -4$

Example 4. Solve this expression for $x = -3$. $15 + 7x$

Solution: Substitute -3 for x.

Then: $15 + 7x = 15 + 7(-3) = 15 - 21 = -6$

Evaluating Two Variables

- To evaluate an algebraic expression, substitute a number for each variable.

- Perform the arithmetic operations to find the value of the expression.

Examples:

Example 1. Calculate this expression for $a = 2$ and $b = -1$. $(4a - 3b)$

Solution: First, substitute 2 for a, and -1 for b.
Then: $4a - 3b = 4(2) - 3(-1)$
Now, use order of operation to find the answer: $4(2) - 3(-1) = 8 + 3 = 11$

Example 2. Evaluate this expression for $x = -2$ and $y = 2$. $(3x + 6y)$

Solution: Substitute -2 for x, and 2 for y.
Then: $3x + 6y = 3(-2) + 6(2) = -6 + 12 = 6$

Example 3. Find the value of this expression $2(6a - 5b)$, when $a = -1$ and $b = 4$.

Solution: Substitute -1 for a, and 4 for b.
Then: $2(6a - 5b) = 2\big(6(-1) - 5(4)\big) = 2(-6 - 20) = 2(-26) = -52$

Example 4. Evaluate this expression. $-7x - 2y$, $x = 4$, $y = -3$

Solution: Substitute 4 for x, and -3 for y and simplify.
Then: $-7x - 2y = -7(4) - 2(-3) = -28 + 6 = -22$

Chapter 7: Practices

✍ Simplify each expression.

1) $(3 + 4x - 1) =$

2) $(-5 - 2x + 7) =$

3) $(12x - 5x - 4) =$

4) $(-16x + 24x - 9) =$

5) $(6x + 5 - 15x) =$

6) $2 + 5x - 8x - 6 =$

7) $5x + 10 - 3x - 22 =$

8) $-5 - 3x^2 - 6 + 4x =$

9) $-6 + 9x^2 - 3 + x =$

10) $5x^2 + 3x - 10x - 3 =$

11) $4x^2 - 2x - 6x + 5 - 8 =$

12) $3x^2 - 5x - 7x + 2 - 4 =$

13) $9x^2 - x - 5x + 3 - 9 =$

14) $2x^2 - 7x - 3x^2 + 4x + 6 =$

✍ Simplify each polynomial.

15) $5x^2 + 3x^3 - 9x^2 + 2x =$ ---

16) $8x^4 + 2x^5 - 7x^4 + 3x^2 =$ --

17) $15x^3 + 11x - 5x^2 - 9x^3 =$ ---

18) $(7x^3 - 3x^2) + (5x^2 - 13x) =$ ---

19) $(12x^4 + 6x^3) + (x^3 - 5x^4) =$ --

20) $(15x^5 - 8x^3) - (4x^3 + x^2) =$ ---

21) $(14x^4 + 7x^3) - (x^3 - 24) =$ ---

22) $(20x^4 + 6x^3) - (-x^3 - 2x^4) =$ ---

23) $(x^2 + 9x^3) + (-22x^2 + 6x^3) =$ ---

24) $(4x^4 - 2x^3) + (-5x^3 - 8x^4) =$ ---

✒ Use the distributive property to simply each expression.

25) $2(6 + x) =$ _____

26) $5(3 - 2x) =$ _____

27) $7(1 - 5x) =$ _____

28) $(3 - 4x)7 =$ _____

29) $6(2 - 3x) =$ _____

30) $(-1)(-9 + x) =$ _____

31) $(-6)(3x - 2) =$ _____

32) $(-x + 12)(-4) =$ _____

33) $(-2)(1 - 6x) =$ _____

34) $(-5x - 3)(-8) =$ _____

✒ Evaluate each expression using the value given.

35) $x = 4 \rightarrow 10 - x =$ ____

36) $x = 6 \rightarrow x + 8 =$ ____

37) $x = 3 \rightarrow 2x - 6 =$ ____

38) $x = 2 \rightarrow 10 - 4x =$ ____

39) $x = 7 \rightarrow 8x - 3 =$ ____

40) $x = 9 \rightarrow 20 - 2x =$ ____

41) $x = 5 \rightarrow 10x - 30 =$ ___

42) $x = -6 \rightarrow 5 - x =$ ____

43) $x = -3 \rightarrow 22 - 3x =$ ____

44) $x = -7 \rightarrow 10 - 9x =$ ____

45) $x = -10 \rightarrow 40 - 3x =$ ____

46) $x = -2 \rightarrow 20x - 5 =$ ____

47) $x = -5 \rightarrow -10x - 8 =$ ___

48) $x = -4 \rightarrow -1 - 4x =$ ___

✒ Evaluate each expression using the values given.

49) $x = 2, y = 1 \rightarrow 2x + 7y =$ _____

50) $a = 3, b = 5 \rightarrow 3a - 5b =$ _____

51) $x = 6, y = 2 \rightarrow 3x - 2y + 8 =$ _____

52) $a = -2, b = 3 \rightarrow -5a + 2b + 6 =$ _____

53) $x = -4, y = -3 \rightarrow -4x + 10 - 8y =$ _____

Chapter 7: Answers

1) $4x + 2$

2) $-2x + 2$

3) $7x - 4$

4) $8x - 9$

5) $-9x + 5$

6) $-3x - 4$

7) $2x - 12$

8) $-3x^2 + 4x - 11$

9) $9x^2 + x - 9$

10) $5x^2 - 7x - 3$

11) $4x^2 - 8x - 3$

12) $3x^2 - 12x - 2$

13) $9x^2 - 6x - 6$

14) $-x^2 - 3x + 6$

15) $3x^3 - 4x^2 + 2x$

16) $2x^5 + x^4 + 3x^2$

17) $6x^3 - 5x^2 + 11x$

18) $7x^3 + 2x^2 - 13x$

19) $7x^4 + 7x^3$

20) $15x^5 - 12x^3 - x^2$

21) $14x^4 + 6x^3 + 24$

22) $22x^4 + 7x^3$

23) $15x^3 - 21x^2$

24) $-4x^4 - 7x^3$

25) $2x + 12$

26) $-10x + 15$

27) $-35x + 7$

28) $-28x + 21$

29) $-18x + 12$

30) $-x + 9$

31) $-18x + 12$

32) $4x - 48$

33) $12x - 2$

34) $40x + 24$

35) 6

36) 14

37) 0

38) 2

39) 53

40) 2

41) 20

42) 11

43) 31

44) 73

45) 70

46) -45

47) 42

48) 15

49) 11

50) -16

51) 22

52) 22

53) 50

CHAPTER

8 Equations and Inequalities

Math topics that you'll learn in this chapter:

☑ One-Step Equations

☑ Multi-Step Equations

☑ System of Equations

☑ Graphing Single–Variable Inequalities

☑ One-Step Inequalities

☑ Multi-Step Inequalities

67

One–Step Equations

- The values of two expressions on both sides of an equation are equal. Example: $ax = b$. In this equation, ax is equal to b.

- Solving an equation means finding the value of the variable.

- You only need to perform one Math operation to solve the one-step equations.

- To solve a one-step equation, find the inverse (opposite) operation is being performed.

- The inverse operations are:

    ❖ Addition and subtraction

    ❖ Multiplication and division

Examples:

Example 1. Solve this equation for x. $4x = 16 \rightarrow x = ?$

Solution: Here, the operation is multiplication (variable x is multiplied by 4) and its inverse operation is division. To solve this equation, divide both sides of equation by 4: $4x = 16 \rightarrow \frac{4x}{4} = \frac{16}{4} \rightarrow x = 4$

Example 2. Solve this equation. $x + 8 = 0 \rightarrow x = ?$

Solution: In this equation, 8 is added to the variable x. The inverse operation of addition is subtraction. To solve this equation, subtract 8 from both sides of the equation: $x + 8 - 8 = 0 - 8$. Then: $x + 8 - 8 = 0 - 8 \rightarrow x = -8$

Example 3. Solve this equation for x. $x - 12 = 0$

Solution: Here, the operation is subtraction and its inverse operation is addition. To solve this equation, add 12 to both sides of the equation: $x - 12 + 12 = 0 + 12 \rightarrow x = 12$

Multi–Step Equations

- To solve a multi-step equation, combine "like" terms on one side.

- Bring variables to one side by adding or subtracting.

- Simplify using the inverse of addition or subtraction.

- Simplify further by using the inverse of multiplication or division.

- Check your solution by plugging the value of the variable into the original equation.

Examples:

Example 1. Solve this equation for x. $4x + 8 = 20 - 2x$

Solution: First, bring variables to one side by adding $2x$ to both sides. Then:

$4x + 8 + 2x = 20 - 2x + 2x \rightarrow 4x + 8 + 2x = 20$.

Simplify: $6x + 8 = 20$. Now, subtract 8 from both sides of the equation:

$6x + 8 - 8 = 20 - 8 \rightarrow 6x = 12 \rightarrow$ Divide both sides by 6:

$6x = 12 \rightarrow \dfrac{6x}{6} = \dfrac{12}{6} \rightarrow x = 2$

Let's check this solution by substituting the value of 2 for x in the original equation:

$x = 2 \rightarrow 4x + 8 = 20 - 2x \rightarrow 4(2) + 8 = 20 - 2(2) \rightarrow 16 = 16$

The answer $x = 2$ is correct.

Example 2. Solve this equation for x. $-5x + 4 = 24$

Solution: Subtract 4 from both sides of the equation.

$-5x + 4 = 24 \rightarrow -5x + 4 - 4 = 24 - 4 \rightarrow -5x = 20$

Divide both sides by -5, then: $-5x = 20 \rightarrow \dfrac{-5x}{-5} = \dfrac{20}{-5} \rightarrow x = -4$

Now, check the solution:

$x = -4 \rightarrow -5x + 4 = 24 \rightarrow -5(-4) + 4 = 24 \rightarrow 24 = 24$

The answer $x = -4$ is correct.

System of Equations

- A system of equations contains two equations and two variables. For example, consider the system of equations: $x - y = 1$ and $x + y = 5$

- The easiest way to solve a system of equations is using the elimination method. The elimination method uses the addition property of equality. You can add the same value to each side of an equation.

- For the first equation above, you can add $x + y$ to the left side and 5 to the right side of the first equation: $x - y + (x + y) = 1 + 5$. Now, if you simplify, you get: $x - y + (x + y) = 1 + 5 \rightarrow 2x = 6 \rightarrow x = 3$. Now, substitute 3 for the x in the first equation: $3 - y = 1$. By solving this equation, $y = 2$

Example:

What is the value of $x + y$ in this system of equations?

$$\begin{cases} 2x + 4y = 12 \\ 4x - 2y = -16 \end{cases}$$

Solution: Solving a System of Equations by Elimination:
Multiply the first equation by (-2), then add it to the second equation.

$$\begin{matrix} -2(2x + 4y = 12) \\ 4x - 2y = -16 \end{matrix} \Rightarrow \begin{matrix} -4x - 8y = -24 \\ 4x - 2y = -16 \end{matrix} \Rightarrow (-4x) + 4x - 8y - 2y = -24 - 16 \Rightarrow$$

$$-10y = -40 \Rightarrow y = 4$$

Plug in the value of y into one of the equations and solve for x.
$2x + 4(4) = 12 \Rightarrow 2x + 16 = 12 \Rightarrow 2x = -4 \Rightarrow x = -2$
Thus, $x + y = -2 + 4 = 2$

Graphing Single–Variable Inequalities

- An inequality compares two expressions using an inequality sign.

- Inequality signs are: "less than" <, "greater than" >, "less than or equal to" ≤, and "greater than or equal to" ≥.

- To graph a single–variable inequality, find the value of the inequality on the number line.

- For less than (<) or greater than (>) draw an open circle on the value of the variable. If there is an equal sign too, then use a filled circle.

- Draw an arrow to the right for greater or to the left for less than.

Examples:

Example 1. Draw a graph for this inequality. $x > 2$

Solution: Since the variable is greater than 2, then we need to find 2 in the number line and draw an open circle on it. Then, draw an arrow to the right.

Example 2. Graph this inequality. $x \leq -3$.

Solution: Since the variable is less than or equal to -3, then we need to find -3 on the number line and draw a filled circle on it. Then, draw an arrow to the left.

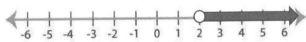

One–Step Inequalities

- An inequality compares two expressions using an inequality sign.
- Inequality signs are: "less than" <, "greater than" >, "less than or equal to" ≤, and "greater than or equal to" ≥.
- You only need to perform one Math operation to solve the one-step inequalities.
- To solve one-step inequalities, find the inverse (opposite) operation is being performed.
- For dividing or multiplying both sides by negative numbers, flip the direction of the inequality sign.

Examples:

Example 1. Solve this inequality for x. $x + 5 \geq 4$

Solution: The inverse (opposite) operation of addition is subtraction. In this inequality, 5 is added to x. To isolate x we need to subtract 5 from both sides of the inequality.
Then: $x + 5 \geq 4 \rightarrow x + 5 - 5 \geq 4 - 5 \rightarrow x \geq -1$. The solution is: $x \geq -1$

Example 2. Solve the inequality. $x - 3 > -6$

Solution: 3 is subtracted from x. Add 3 to both sides.
$x - 3 > -6 \rightarrow x - 3 + 3 > -6 + 3 \rightarrow x > -3$

Example 3. Solve. $4x \leq -8$

Solution: 4 is multiplied to x. Divide both sides by 4.
Then: $4x \leq -8 \rightarrow \frac{4x}{4} \leq \frac{-8}{4} \rightarrow x \leq -2$

Example 4. Solve. $-3x \leq 6$

Solution: -3 is multiplied to x. Divide both sides by -3. Remember when dividing or multiplying both sides of an inequality by negative numbers, flip the direction of the inequality sign.
Then: $-3x \leq 6 \rightarrow \frac{-3x}{-3} \geq \frac{6}{-3} \rightarrow x \geq -2$

www.EffortlessMath.com

Multi–Step Inequalities

- To solve a multi-step inequality, combine "like" terms on one side.

- Bring variables to one side by adding or subtracting.

- Isolate the variable.

- Simplify using the inverse of addition or subtraction.

- Simplify further by using the inverse of multiplication or division.

- For dividing or multiplying both sides by negative numbers, flip the direction of the inequality sign.

Examples:

Example 1. Solve this inequality. $8x - 2 \leq 14$

Solution: In this inequality, 2 is subtracted from $8x$. The inverse of subtraction is addition. Add 2 to both sides of the inequality:

$8x - 2 + 2 \leq 14 + 2 \rightarrow 8x \leq 16$

Now, divide both sides by 8. Then: $8x \leq 16 \rightarrow \frac{8x}{8} \leq \frac{16}{8} \rightarrow x \leq 2$

The solution of this inequality is $x \leq 2$.

Example 2. Solve this inequality. $3x + 9 < 12$

Solution: First, subtract 9 from both sides: $3x + 9 - 9 < 12 - 9$

Then simplify: $3x + 9 - 9 < 12 - 9 \rightarrow 3x < 3$

Now divide both sides by 3: $\frac{3x}{3} < \frac{3}{3} \rightarrow x < 1$

Example 3. Solve this inequality. $-5x + 3 \geq 8$

Solution: First, subtract 3 from both sides:

$-5x + 3 - 3 \geq 8 - 3 \rightarrow -5x \geq 5$

Divide both sides by -5. Remember that you need to flip the direction of inequality sign. $-5x \geq 5 \rightarrow \frac{-5x}{-5} \leq \frac{5}{-5} \rightarrow x \leq -1$

Chapter 8: Practices

✍ Solve each equation. (One−Step Equations)

1) $x + 6 = 3 \to x =$ ____

2) $5 = 11 - x \to x =$ ____

3) $-3 = 8 + x \to x =$ ____

4) $x - 2 = -7 \to x =$ ____

5) $-15 = x + 6 \to x =$ ____

6) $10 - x = -2 \to x =$ ____

7) $22 - x = -9 \to x =$ ____

8) $-4 + x = 28 \to x =$ ____

9) $11 - x = -7 \to x =$ ____

10) $35 - x = -7 \to x =$ ____

✍ Solve each equation. (Multi−Step Equations)

11) $4(x + 2) = 12 \to x =$ ____

12) $-6(6 - x) = 12 \to x =$ ____

13) $5 = -5(x + 2) \to x =$ ____

14) $-10 = 2(4 + x) \to x =$ ____

15) $4(x + 2) = -12, x =$ ____

16) $-6(3 + 2x) = 30, x =$ ____

17) $-3(4 - x) = 12, x =$ ____

18) $-4(6 - x) = 16, x =$ ____

✍ Solve each system of equations.

19) $\begin{cases} x + 6y = 32 \\ x + 3y = 17 \end{cases}$ $x =$ ____ $y =$ ____

20) $\begin{cases} 3x + y = 15 \\ x + 2y = 10 \end{cases}$ $x =$ ____ $y =$ ____

21) $\begin{cases} 3x + 5y = 17 \\ 2x + y = 9 \end{cases}$ $x =$ ____ $y =$ ____

22) $\begin{cases} 5x - 2y = -8 \\ -6x + 2y = 10 \end{cases}$ $x =$ ____ $y =$ ____

✎ **Draw a graph for each inequality.**

23) $x \leq -3$

24) $x > -5$

✎ **Solve each inequality and graph it.**

25) $x - 2 \geq -2$

26) $2x - 3 < 9$

✎ **Solve each inequality.**

27) $x + 13 > 4$

28) $x + 6 > 5$

29) $-12 + 2x \leq 26$

30) $-2 + 8x \leq 14$

31) $6 + 4x \leq 18$

32) $4(x + 3) \geq -12$

33) $2(6 + x) \geq -12$

34) $3(x - 5) < -6$

35) $10 + 5x < -15$

36) $6(6 + x) \geq -18$

37) $2(x - 5) \geq -14$

38) $6(x + 4) < -12$

39) $3(x - 8) \geq -48$

40) $-(6 - 4x) > -30$

41) $2(2 + 2x) > -60$

42) $-3(4 + 2x) > -24$

Chapter 8: Answers

1) -3

2) 6

3) -11

4) -5

5) -21

6) 12

7) 31

8) 32

9) 18

10) 42

11) 1

12) 8

13) -3

14) -9

15) -5

16) -4

17) 8

18) 10

19) $x = 2, y = 5$

20) $x = 4, y = 3$

21) $x = 4, y = 1$

22) $x = -2, y = -1$

23) $x \leq -3$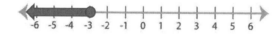

24) $x > -5$

25) $x \geq 0$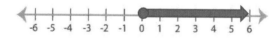

26) $x < 6$

27) $x > -9$

28) $x > -1$

29) $x \leq 19$

30) $x \leq 2$

31) $x \leq 3$

32) $x \geq -6$

33) $x \geq -12$

34) $x < 3$

35) $x < -5$

36) $x \geq -9$

37) $x \geq -2$

38) $x < -6$

39) $x \geq -8$

40) $x > -6$

41) $x > -16$

42) $x < 2$

CHAPTER

9 Lines and Slope

Math topics that you'll learn in this chapter:

- ☑ Finding Slope
- ☑ Graphing Lines Using Slope–Intercept Form
- ☑ Writing Linear Equations
- ☑ Finding Midpoint
- ☑ Finding Distance of Two Points
- ☑ Graphing Linear Inequalities

77

Finding Slope

- The slope of a line represents the direction of a line on the coordinate plane.

- A coordinate plane contains two perpendicular number lines. The horizontal line is x and the vertical line is y. The point at which the two axes intersect is called the origin. An ordered pair (x, y) shows the location of a point.

- A line on a coordinate plane can be drawn by connecting two points.

- To find the slope of a line, we need the equation of the line or two points on the line.

- The slope of a line with two points A (x_1, y_1) and B (x_2, y_2) can be found by using this formula: $\frac{y_2 - y_1}{x_2 - x_1} = \frac{rise}{run}$

- The equation of a line is typically written as $y = mx + b$ where m is the slope and b is the y-intercept.

Examples:

Example 1. Find the slope of the line through these two points:

$$A(1, -6) \ and \ B(3, 2).$$

Solution: Slope $= \frac{y_2 - y_1}{x_2 - x_1}$. Let (x_1, y_1) be A$(1, -6)$ and (x_2, y_2) be $B(3, 2)$.

(Remember, you can choose any point for (x_1, y_1) and (x_2, y_2)).

Then: slope $= \frac{y_2 - y_1}{x_2 - x_1} = \frac{2-(-6)}{3-1} = \frac{8}{2} = 4$

The slope of the line through these two points is 4.

Example 2. Find the slope of the line with equation $y = -2x + 8$

Solution: When the equation of a line is written in the form of $y = mx + b$, the slope is m. In this line: $y = -2x + 8$, the slope is -2.

Graphing Lines Using Slope–Intercept Form

- Slope–intercept form of a line: given the slope **m** and the **y**–intercept (the intersection of the line and *y*-axis) **b**, then the equation of the line is:

$$y = mx + b$$

- To draw the graph of a linear equation in a slope-intercept form on the *xy* coordinate plane, find two points on the line by plugging two values for *x* and calculating the values of *y*.

- You can also use the slope (*m*) and one point to graph the line.

Example:

Sketch the graph of $y = 2x - 4$.

Solution: To graph this line, we need to find two points. When *x* is zero the value of *y* is −4. And when *x* is 2 the value of *y* is 0.

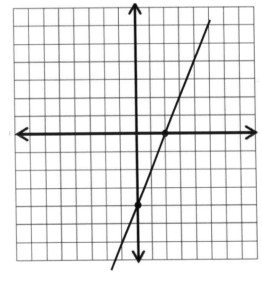

$$x = 0 \rightarrow y = 2(0) - 4 = -4,$$
$$y = 0 \rightarrow 0 = 2x - 4 \rightarrow x = 2$$

Now, we have two points:
$(0, -4)$ and $(2, 0)$.
Find the points on the coordinate plane and graph the line. Remember that the slope of the line is 2.

Writing Linear Equations

- The equation of a line in slope-intercept form: $y = mx + b$

- To write the equation of a line, first identify the slope.

- Find the y-intercept. This can be done by substituting the slope and the coordinates of a point (x, y) on the line.

Examples:

Example 1. What is the equation of the line that passes through $(3, -4)$ and has a slope of 6?

Solution: The general slope-intercept form of the equation of a line is $y = mx + b$, where m is the slope and b is the y-intercept.
By substitution of the given point and given slope:
$y = mx + b \rightarrow -4 = (6)(3) + b$. So, $b = -4 - 18 = -22$, and the required equation of the line is: $y = 6x - 22$

Example 2. Write the equation of the line through two points $A(3, 1)$ and $B(-2, 6)$.

Solution: First, find the slope: $Slop = \frac{y_2 - y_1}{x_2 - x_1} = \frac{6 - 1}{-2 - 3} = \frac{5}{-5} = -1 \rightarrow m = -1$
To find the value of b, use either points and plug in the values of x and y in the equation. The answer will be the same: $y = -x + b$. Let's check both points.
Then: $(3, 1) \rightarrow y = mx + b \rightarrow 1 = -1(3) + b \rightarrow b = 4$
$(-2, 6) \rightarrow y = mx + b \rightarrow 6 = -1(-2) + b \rightarrow b = 4$.
The y-intercept of the line is 4. The equation of the line is: $y = -x + 4$

Example 3. What is the equation of the line that passes through $(4, -1)$ and has a slope of 4?

Solution: The general slope-intercept form of the equation of a line is $y = mx + b$, where m is the slope and b is the y-intercept. By substitution of the given point and given slope: $y = mx + b \rightarrow -1 = (4)(4) + b$
So, $b = -1 - 16 = -17$, and the equation of the line is: $y = 4x - 17$.

Finding Midpoint

- The middle of a line segment is its midpoint.

- The Midpoint of two endpoints A (x_1, y_1) and B (x_2, y_2) can be found using this formula: M $(\frac{x_1+x_2}{2}, \frac{y_1+y_2}{2})$

Examples:

Example 1. Find the midpoint of the line segment with the given endpoints. $(2, -4), (6, 8)$

Solution: Midpoint $= \left(\frac{x_1+x_2}{2}, \frac{y_1+y_2}{2}\right) \rightarrow (x_1, y_1) = (2, -4)$ and $(x_2, y_2) = (6, 8)$

Midpoint $= \left(\frac{2+6}{2}, \frac{-4+8}{2}\right) \rightarrow \left(\frac{8}{2}, \frac{4}{2}\right) \rightarrow M(4, 2)$

Example 2. Find the midpoint of the line segment with the given endpoints. $(-2, 3), (6, -7)$

Solution: Midpoint $= \left(\frac{x_1+x_2}{2}, \frac{y_1+y_2}{2}\right) \rightarrow (x_1, y_1) = (-2, 3)$ and $(x_2, y_2) = (6, -7)$

Midpoint $= \left(\frac{-2+6}{2}, \frac{3+(-7)}{2}\right) \rightarrow \left(\frac{4}{2}, \frac{-4}{2}\right) \rightarrow M(2, -2)$

Example 3. Find the midpoint of the line segment with the given endpoints. $(7, -4), (1, 8)$

Solution: Midpoint $= \left(\frac{x_1+x_2}{2}, \frac{y_1+y_2}{2}\right) \rightarrow (x_1, y_1) = (7, -4)$ and $(x_2, y_2) = (1, 8)$

Midpoint $= \left(\frac{7+1}{2}, \frac{-4+8}{2}\right) \rightarrow \left(\frac{8}{2}, \frac{4}{2}\right) \rightarrow M(4, 2)$

Example 4. Find the midpoint of the line segment with the given endpoints. $(6, 3), (10, -9)$

Solution: Midpoint $= \left(\frac{x_1+x_2}{2}, \frac{y_1+y_2}{2}\right) \rightarrow (x_1, y_1) = (6, 3)$ and $(x_2, y_2) = (10, -9)$

Midpoint $= \left(\frac{6+10}{2}, \frac{3-9}{2}\right) \rightarrow \left(\frac{16}{2}, \frac{-6}{2}\right) \rightarrow M(8, -3)$

Finding Distance of Two Points

- Use the following formula to find the distance of two points with the coordinates A (x_1, y_1) and B (x_2, y_2):

$$d = \sqrt{(x_2 - x_1)^2 + (y_2 - y_1)^2}$$

Examples:

Example 1. Find the distance between $(4, 2)$ and $(-5, -10)$ on the coordinate plane.

Solution: Use distance of two points formula: $d = \sqrt{(x_2 - x_1)^2 + (y_2 - y_1)^2}$

$(x_1, y_1) = (4, 2)$ and $(x_2, y_2) = (-5, -10)$. Then: $d = \sqrt{(x_2 - x_1)^2 + (y_2 - y_1)^2} \rightarrow$

$= \sqrt{(-5 - 4)^2 + (-10 - 2)^2} = \sqrt{(-9)^2 + (-12)^2} = \sqrt{81 + 144} = \sqrt{225} = 15$

Then: $d = 15$

Example 2. Find the distance of two points $(-1, 5)$ and $(-4, 1)$.

Solution: Use distance of two points formula: $d = \sqrt{(x_2 - x_1)^2 + (y_2 - y_1)^2}$

$(x_1, y_1) = (-1, 5)$, and $(x_2, y_2) = (-4, 1)$

Then: $= \sqrt{(x_2 - x_1)^2 + (y_2 - y_1)^2} \rightarrow d = \sqrt{(-4 - (-1))^2 + (1 - 5)^2} =$

$\sqrt{(-3)^2 + (-4)^2} = \sqrt{9 + 16} = \sqrt{25} = 5$. Then: $d = 5$

Example 3. Find the distance between $(-6, 5)$ and $(-1, -7)$.

Solution: Use distance of two points formula: $d = \sqrt{(x_2 - x_1)^2 + (y_2 - y_1)^2}$

$(x_1, y_1) = (-6, 5)$ and $(x_2, y_2) = (-1, -7)$. Then: $d = \sqrt{(x_2 - x_1)^2 + (y_2 - y_1)^2}$

$d = \sqrt{(-1 - (-6))^2 + (-7 - 5)^2} = \sqrt{(5)^2 + (-12)^2} = \sqrt{25 + 144} = \sqrt{169} = 13$

Graphing Linear Inequalities

- To graph a linear inequality, first draw a graph of the "equals" line.

- Use a dash line for less than (<) and greater than (>) signs and a solid line for less than and equal to (≤) and greater than and equal to (≥).

- Choose a testing point. (it can be any point on both sides of the line.)

- Put the value of (x, y) of that point in the inequality. If that works, that part of the line is the solution. If the values don't work, then the other part of the line is the solution.

Example:

Sketch the graph of inequality: $y < 2x + 4$

Solution: To draw the graph of $y < 2x + 4$, you first need to graph the line:

$y = 2x + 4$

Since there is a less than (<) sign, draw a dash line.

The slope is 2 and y-intercept is 4.

Then, choose a testing point and substitute the value of x and y from that point into the inequality. The easiest point to test is the origin: $(0, 0)$

$$(0,0) \rightarrow y < 2x + 4 \rightarrow 0 < 2(0) + 4 \rightarrow 0 < 4$$

This is correct! 0 is less than 4. So, this part of the line (on the right side) is the solution of this inequality.

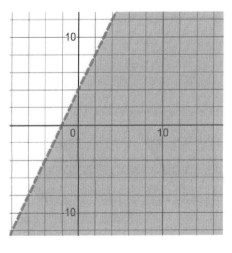

Chapter 9: Practices

✍ Find the slope of each line.

1) $y = x - 5$

2) $y = 2x + 6$

3) $y = -5x - 8$

4) Line through $(2, 6)$ *and* $(5, 0)$

5) Line through $(8, 0)$ *and* $(-4, 3)$

6) Line through $(-2, -4)$ *and* $(-4, 8)$

✍ Sketch the graph of each line. (Using Slope−Intercept Form)

7) $y = x + 4$

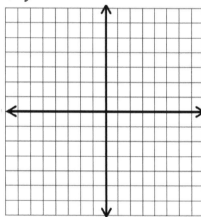

8) $y = 2x - 5$

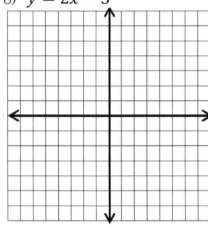

✍ Solve.

9) What is the equation of a line with slope 4 and intercept 16? _____

10) What is the equation of a line with slope 3 and passes through point $(1, 5)$?

_____

11) What is the equation of a line with slope -5 and passes through point $(-2, 7)$?

_____

12) The slope of a line is -4 and it passes through point $(-6, 2)$. What is the equation of the line? _____

13) The slope of a line is -3 and it passes through point $(-3, -6)$. What is the equation of the line? _____

✎ Sketch the graph of each linear inequality.

14) $y > 2x - 2$

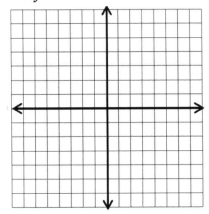

15) $y < -x + 3$

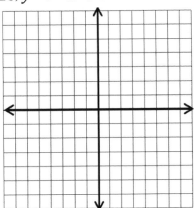

✎ Find the midpoint of the line segment with the given endpoints.

16) $(5, 0), (1, 4)$

17) $(2, 3), (4, 7)$

18) $(8, 1), (2, 5)$

19) $(5, 10), (3, 6)$

20) $(4, -1), (-2, 7)$

21) $(2, -5), (4, 1)$

22) $(7, 6), (-5, 2)$

23) $(-2, 8), (4, -6)$

✎ Find the distance between each pair of points.

24) $(-2, 8), (-6, 8)$

25) $(4, -4), (14, 20)$

26) $(-1, 9), (-5, 6)$

27) $(0, 3), (4, 3)$

28) $(0, -2), (5, 10)$

29) $(4, 3), (7, -1)$

30) $(2, 6), (10, -9)$

31) $(3, 3), (6, -1)$

32) $(-2, -12), (14, 18)$

33) $(2, -2), (12, 22)$

Chapter 9: Answers

1) 1   3) -5 5) $-\frac{1}{4}$

2) 2   4) -2 6) -6

7) $y = x + 4$ 8) $y = 2x - 5$

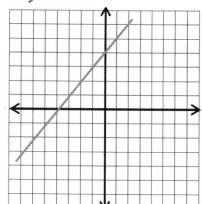   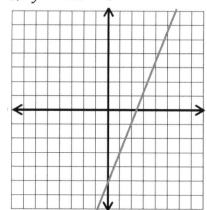

9) $y = 4x + 16$ 11) $y = -5x - 3$ 13) $y = -3x - 15$

10) $y = 3x + 2$ 12) $y = -4x - 22$

14) $y > 2x - 2$ 15) $y < -x + 3$

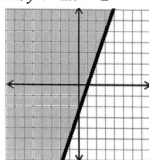   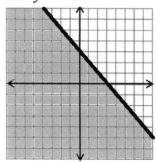

16) $(3, 2)$ 22) $(1, 4)$ 28) 13

17) $(3, 5)$ 23) $(1, 1)$ 29) 5

18) $(5, 3)$ 24) 4   30) 17

19) $(4, 8)$ 25) 26   31) 5

20) $(1, 3)$ 26) 5   32) 34

21) $(3, -2)$ 27) 4   33) 26

CHAPTER

10 Polynomials

Math topics that you'll learn in this chapter:

- ☑ Simplifying Polynomials
- ☑ Adding and Subtracting Polynomials
- ☑ Multiplying Monomials
- ☑ Multiplying and Dividing Monomials
- ☑ Multiplying a Polynomial and a Monomial
- ☑ Multiplying Binomials
- ☑ Factoring Trinomials

87

Simplifying Polynomials

- To simplify Polynomials, find "like" terms. (they have same variables with same power).

- Use "FOIL". (First–Out–In–Last) for binomials:

$$(x + a)(x + b) = x^2 + (b + a)x + ab$$

- Add or Subtract "like" terms using order of operation.

Examples:

Example 1. Simplify this expression. $x(4x + 7) - 2x =$

Solution: Use Distributive Property: $x(4x + 7) = 4x^2 + 7x$
Now, combine like terms: $x(4x + 7) - 2x = 4x^2 + 7x - 2x = 4x^2 + 5x$

Example 2. Simplify this expression. $(x + 3)(x + 5) =$

Solution: First, apply the FOIL method: $(a + b)(c + d) = ac + ad + bc + bd$
$(x + 3)(x + 5) = x^2 + 5x + 3x + 15$
Now combine like terms: $x^2 + 5x + 3x + 15 = x^2 + 8x + 15$

Example 3. Simplify this expression. $2x(x - 5) - 3x^2 + 6x =$

Solution: Use Distributive Property: $2x(x - 5) = 2x^2 - 10x$
Then: $2x(x - 5) - 3x^2 + 6x = 2x^2 - 10x - 3x^2 + 6x$
Now combine like terms: $2x^2 - 3x^2 = -x^2$, and $-10x + 6x = -4x$
The simplified form of the expression: $2x^2 - 10x - 3x^2 + 6 = -x^2 - 4x$

Adding and Subtracting Polynomials

- Adding polynomials is just a matter of combining like terms, with some order of operations considerations thrown in.

- Be careful with the minus signs, and don't confuse addition and multiplication!

- For subtracting polynomials, sometimes you need to use the Distributive Property: $a(b + c) = ab + ac$, $a(b - c) = ab - ac$

Examples:

Example 1. Simplify the expressions. $(x^2 - 2x^3) - (x^3 - 3x^2) =$

Solution: First, use Distributive Property:
$-(x^3 - 3x^2) = -x^3 + 3x^2$
$\rightarrow (x^2 - 2x^3) - (x^3 - 3x^2) = x^2 - 2x^3 - x^3 + 3x^2$
Now combine like terms: $-2x^3 - x^3 = -3x^3$ and $x^2 + 3x^2 = 4x^2$
Then: $(x^2 - 2x^3) - (x^3 - 3x^2) = x^2 - 2x^3 - x^3 + 3x^2 = -3x^3 + 4x^2$

Example 2. Add expressions. $(3x^3 - 5) + (4x^3 - 2x^2) =$

Solution: Remove parentheses:
$$(3x^3 - 5) + (4x^3 - 2x^2) = 3x^3 - 5 + 4x^3 - 2x^2$$
Now combine like terms: $3x^3 - 5 + 4x^3 - 2x^2 = 7x^3 - 2x^2 - 5$

Example 3. Simplify the expressions. $(-4x^2 - 2x^3) - (5x^2 + 2x^3) =$

Solution: First, use Distributive Property: $-(5x^2 + 2x^3) = -5x^2 - 2x^3 \rightarrow$
$$(-4x^2 - 2x^3) - (5x^2 + 2x^3) = -4x^2 - 2x^3 - 5x^2 - 2x^3$$
Now combine like terms and write in standard form:
$-4x^2 - 2x^3 - 5x^2 - 2x^3 = -4x^3 - 9x^2$

Multiplying Monomials

- A monomial is a polynomial with just one term: Examples: $2x$ or $7y^2$.

- When you multiply monomials, first multiply the coefficients (a number placed before and multiplying the variable) and then multiply the variables using multiplication property of exponents.

$$x^a \times x^b = x^{a+b}$$

Examples:

Example 1. Multiply expressions. $2xy^3 \times 6x^4y^2$

Solution: Find the same variables and use multiplication property of exponents: $x^a \times x^b = x^{a+b}$
$x \times x^4 = x^{1+4} = x^5$ and $y^3 \times y^2 = y^{3+2} = y^5$
Then, multiply coefficients and variables: $2xy^3 \times 6x^4y^2 = 12x^5y^5$

Example 2. Multiply expressions. $7a^3b^8 \times 3a^6b^4 =$

Solution: Use the multiplication property of exponents: $x^a \times x^b = x^{a+b}$
$a^3 \times a^6 = a^{3+6} = a^9$ and $b^8 \times b^4 = b^{8+4} = b^{12}$
Then: $7a^3b^8 \times 3a^6b^4 = 21a^9b^{12}$

Example 3. Multiply. $5x^2y^4z^3 \times 4x^4y^7z^5$

Solution: Use the multiplication property of exponents: $x^a \times x^b = x^{a+b}$
$x^2 \times x^4 = x^{2+4} = x^6$, $y^4 \times y^7 = y^{4+7} = y^{11}$ and $z^3 \times z^5 = z^{3+5} = z^8$
Then: $5x^2y^4z^3 \times 4x^4y^7z^5 = 20x^6y^{11}z^8$

Example 4. Simplify. $(-6a^7b^4)(4a^8b^5) =$

Solution: Use the multiplication property of exponents: $x^a \times x^b = x^{a+b}$
$a^7 \times a^8 = a^{7+8} = a^{15}$ and $b^4 \times b^5 = b^{4+5} = b^9$
Then: $(-6a^7b^4)(4a^8b^5) = -24a^{15}b^9$

Multiplying and Dividing Monomials

- When you divide or multiply two monomials, you need to divide or multiply their coefficients and then divide or multiply their variables.

- In case of exponents with the same base, for Division, subtract their powers, for Multiplication, add their powers.

- Exponent's Multiplication and Division rules:

$$x^a \times x^b = x^{a+b}, \qquad \frac{x^a}{x^b} = x^{a-b}$$

Examples:

Example 1. Multiply expressions. $(3x^5)(9x^4) =$

Solution: Use multiplication property of exponents:
$x^a \times x^b = x^{a+b} \rightarrow x^5 \times x^4 = x^9$
Then: $(3x^5)(9x^4) = 27x^9$

Example 2. Divide expressions. $\frac{12x^4y^6}{6xy^2} =$

Solution: Use division property of exponents:
$\frac{x^a}{x^b} = x^{a-b} \rightarrow \frac{x^4}{x} = x^{4-1} = x^3$ and $\frac{y^6}{y^2} = y^{6-2} = y^4$
Then: $\frac{12x^4y^6}{6xy^2} = 2x^3y^4$

Example 3. Divide expressions. $\frac{49a^6b^9}{7a^3b^4}$

Solution: Use division property of exponents:
$\frac{x^a}{x^b} = x^{a-b} \rightarrow \frac{a^6}{a^3} = a^{6-3} = a^3$ and $\frac{b^9}{b^4} = b^{9-4} = b^5$
Then: $\frac{49a^6b^9}{7a^3b^4} = 7a^3b^5$

Multiplying a Polynomial and a Monomial

- When multiplying monomials, use the product rule for exponents.

$$x^a \times x^b = x^{a+b}$$

- When multiplying a monomial by a polynomial, use the distributive property.

$$a \times (b + c) = a \times b + a \times c = ab + ac$$
$$a \times (b - c) = a \times b - a \times c = ab - ac$$

Examples:

Example 1. Multiply expressions. $6x(2x + 5)$

Solution: Use Distributive Property:
$6x(2x + 5) = 6x \times 2x + 6x \times 5 = 12x^2 + 30x$

Example 2. Multiply expressions. $x(3x^2 + 4y^2)$

Solution: Use Distributive Property:
$x(3x^2 + 4y^2) = x \times 3x^2 + x \times 4y^2 = 3x^3 + 4xy^2$

Example 3. Multiply. $-x(-2x^2 + 4x + 5)$

Solution: Use Distributive Property:
$-x(-2x^2 + 4x + 5) = (-x)(-2x^2) + (-x) \times (4x) + (-x) \times (5) =$
Now simplify:
$(-x)(-2x^2) + (-x) \times (4x) + (-x) \times (5) = 2x^3 - 4x^2 - 5x$

Multiplying Binomials

- A binomial is a polynomial that is the sum or the difference of two terms, each of which is a monomial.

- To multiply two binomials, use the "FOIL" method. (First–Out–In–Last)

$$(x + a)(x + b) = x \times x + x \times b + a \times x + a \times b = x^2 + bx + ax + ab$$

Examples:

Example 1. Multiply Binomials. $(x + 3)(x - 2) =$

Solution: Use "FOIL". (First–Out–In–Last):
$(x + 3)(x - 2) = x^2 - 2x + 3x - 6$
Then combine like terms: $x^2 - 2x + 3x - 6 = x^2 + x - 6$

Example 2. Multiply. $(x + 6)(x + 4) =$

Solution: Use "FOIL". (First–Out–In–Last):
$(x + 6)(x + 4) = x^2 + 4x + 6x + 24$
Then simplify: $x^2 + 4x + 6x + 24 = x^2 + 10x + 24$

Example 3. Multiply. $(x + 5)(x - 7) =$

Solution: Use "FOIL". (First–Out–In–Last):
$(x + 5)(x - 7) = x^2 - 7x + 5x - 35$
Then simplify: $x^2 - 7x + 5x - 35 = x^2 - 2x - 35$

Example 4. Multiply Binomials. $(x - 9)(x - 5) =$

Solution: Use "FOIL". (First–Out–In–Last):
$(x - 9)(x - 5) = x^2 - 5x - 9x + 45$
Then combine like terms: $x^2 - 5x - 9x + 45 = x^2 - 14x + 45$

Factoring Trinomials

To factor trinomials, you can use following methods:

- "FOIL": $(x + a)(x + b) = x^2 + (b + a)x + ab$

- "Difference of Squares":

$$a^2 - b^2 = (a + b)(a - b)$$
$$a^2 + 2ab + b^2 = (a + b)(a + b)$$
$$a^2 - 2ab + b^2 = (a - b)(a - b)$$

- "Reverse FOIL": $x^2 + (b + a)x + ab = (x + a)(x + b)$

Examples:

Example 1. Factor this trinomial. $x^2 - 2x - 8$

Solution: Break the expression into groups. You need to find two numbers that their product is -8 and their sum is -2. (remember "Reverse FOIL": $x^2 + (b + a)x + ab = (x + a)(x + b)$). Those two numbers are 2 and -4. Then:
$$x^2 - 2x - 8 = (x^2 + 2x) + (-4x - 8)$$
Now factor out x from $x^2 + 2x : x(x + 2)$, and factor out -4 from $-4x - 8: -4(x + 2)$; Then: $(x^2 + 2x) + (-4x - 8) = x(x + 2) - 4(x + 2)$
Now factor out like term: $(x + 2)$. Then: $(x + 2)(x - 4)$

Example 2. Factor this trinomial. $x^2 - 2x - 24$

Solution: Break the expression into groups: $(x^2 + 4x) + (-6x - 24)$
Now factor out x from $x^2 + 4x : x(x + 4)$, and factor out -6 from $-6x - 24: -6(x + 4)$; Then: $(x + 4) - 6(x + 4)$, now factor out like term: $(x + 4) \rightarrow x(x + 4) - 6(x + 4) = (x + 4)(x - 6)$

Chapter 10: Practices

✎ Simplify each polynomial.

1) $3(6x + 4) =$

2) $5(3x - 8) =$

3) $x(7x + 2) + 9x =$

4) $6x(x + 3) + 5x =$

5) $6x(3x + 1) - 5x =$

6) $x(3x - 4) + 3x^2 - 6 =$

7) $x^2 - 5 - 3x(x + 8) =$

8) $2x^2 + 7 - 6x(2x + 5) =$

✎ Add or subtract polynomials.

9) $(x^2 + 3) + (2x^2 - 4) =$

10) $(3x^2 - 6x) - (x^2 + 8x) =$

11) $(4x^3 - 3x^2) + (2x^3 - 5x^2) =$

12) $(6x^3 - 7x) - (5x^3 - 3x) =$

13) $(10x^3 + 4x^2) + (14x^2 - 8) =$

14) $(4x^3 - 9) - (3x^3 - 7x^2) =$

15) $(9x^3 + 3x) - (6x^3 - 4x) =$

16) $(7x^3 - 5x) - (3x^3 + 5x) =$

✎ Find the products. (Multiplying Monomials)

17) $3x^2 \times 8x^3 =$

18) $2x^4 \times 9x^3 =$

19) $-4a^4b \times 2ab^3 =$

20) $(-7x^3yz) \times (3xy^2z^4) =$

21) $-2a^5bc \times 6a^2b^4 =$

22) $9u^3t^2 \times (-2ut) =$

23) $12x^2z \times 3xy^3 =$

24) $11x^3z \times 5xy^5 =$

25) $-6a^3bc \times 5a^4b^3 =$

26) $-4x^6y^2 \times (-12xy) =$

✎ **Simplify each expression. (Multiplying and Dividing Monomials)**

27) $(7x^2y^3)(3x^4y^2) =$

28) $(6x^3y^2)(4x^4y^3) =$

29) $(10x^8y^5)(3x^5y^7) =$

30) $(15a^3b^2)(2a^3b^8) =$

31) $\frac{42x^4y^2}{6x^3y} =$

32) $\frac{49x^5y^6}{7x^2y} =$

33) $\frac{63x^{15}y^{10}}{9x^8y^6} =$

34) $\frac{35x^8y^{12}}{5x^4y^8} =$

✎ **Find each product. (Multiplying a Polynomial and a Monomial)**

35) $3x(5x - y) =$

36) $2x(4x + y) =$

37) $7x(x - 3y) =$

38) $x(2x^2 + 2x - 4) =$

39) $5x(3x^2 + 8x + 2) =$

40) $7x(2x^2 - 9x - 5) =$

✎ **Find each product. (Multiplying Binomials)**

41) $(x - 3)(x + 3) =$

42) $(x - 6)(x + 6) =$

43) $(x + 10)(x + 4) =$

44) $(x - 6)(x + 7) =$

45) $(x + 2)(x - 5) =$

46) $(x - 10)(x + 3) =$

✎ **Factor each trinomial.**

47) $x^2 + 6x + 8 =$

48) $x^2 + 3x - 10 =$

49) $x^2 + 2x - 48 =$

50) $x^2 - 10x + 80 =$

51) $2x^2 - 7x + 12 =$

52) $3x^2 - 10x + 3 =$

Chapter 10: Answers

1) $18x + 12$

2) $15x - 40$

3) $7x^2 + 11x$

4) $6x^2 + 23x$

5) $18x^2 + x$

6) $6x^2 - 4x - 6$

7) $-2x^2 - 24x - 5$

8) $-10x^2 - 30x + 7$

9) $3x^2 - 1$

10) $2x^2 - 14x$

11) $6x^3 - 8x^2$

12) $x^3 - 4x$

13) $10x^3 + 18x^2 - 8$

14) $x^3 + 7x^2 - 9$

15) $3x^3 + 7x$

16) $4x^3 - 10x$

17) $24x^5$

18) $18x^7$

19) $-8a^5b^4$

20) $-21x^4y^3z^5$

21) $-12a^7b^5c$

22) $-18u^4t^3$

23) $36x^3y^3z$

24) $55x^4y^5z$

25) $-30a^7b^4c$

26) $48x^7y^3$

27) $21x^6y^5$

28) $24x^7y^5$

29) $30x^{13}y^{12}$

30) $30a^6b^{10}$

31) $7xy$

32) $7x^3y^5$

33) $7x^7y^4$

34) $7x^4y^4$

35) $15x^2 - 3xy$

36) $8x^2 + 2xy$

37) $7x^2 - 21xy$

38) $2x^3 + 2x^2 - 4x$

39) $15x^3 + 40x^2 + 10x$

40) $14x^3 - 63x^2 - 35x$

41) $x^2 - 9$

42) $x^2 - 36$

43) $x^2 + 14x + 40$

44) $x^2 + x - 42$

45) $x^2 - 3x - 10$

46) $x^2 - 7x - 30$

47) $(x + 4)(x + 2)$

48) $(x + 5)(x - 2)$

49) $(x - 6)(x + 8)$

50) $(x - 8)(x - 2)$

51) $(2x - 4)(x - 3)$

52) $(3x - 1)(x - 3)$

CHAPTER

11 Geometry and Solid Figures

Math topics that you'll learn in this chapter:

- ☑ The Pythagorean Theorem
- ☑ Complementary and Supplementary angles
- ☑ Parallel lines and Transversals
- ☑ Triangles
- ☑ Special Right Triangles
- ☑ Polygons
- ☑ Circles
- ☑ Trapezoids
- ☑ Cubes
- ☑ Rectangle Prisms
- ☑ Cylinder

99

The Pythagorean Theorem

- You can use the Pythagorean Theorem to find a missing side in a right triangle.

- In any right triangle: $a^2 + b^2 = c^2$

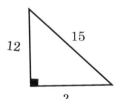

Examples:

Example 1. Right triangle ABC (not shown) has two legs of lengths 3 cm (AB) and 4 cm (AC). What is the length of the hypotenuse of the triangle (side BC)?

Solution: Use Pythagorean Theorem: $a^2 + b^2 = c^2$, $a = 3$, and $b = 4$

Then: $a^2 + b^2 = c^2 \rightarrow 3^2 + 4^2 = c^2 \rightarrow 9 + 16 = c^2 \rightarrow 25 = c^2 \rightarrow c = \sqrt{25} = 5$

The length of the hypotenuse is 5 cm.

Example 2. Find the hypotenuse of this triangle.

Solution: Use Pythagorean Theorem: $a^2 + b^2 = c^2$

Then: $a^2 + b^2 = c^2 \rightarrow 8^2 + 6^2 = c^2 \rightarrow 64 + 36 = c^2$

$c^2 = 100 \rightarrow c = \sqrt{100} = 10$

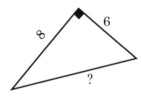

Example 3. Find the length of the missing side in this triangle.

Solution: Use Pythagorean Theorem: $a^2 + b^2 = c^2$

Then: $a^2 + b^2 = c^2 \rightarrow 12^2 + b^2 = 15^2 \rightarrow 144 + b^2 = 225 \rightarrow$

$$b^2 = 225 - 144 \rightarrow b^2 = 81 \rightarrow b = \sqrt{81} = 9$$

Complementary and Supplementary angles

- Two angles with a sum of 90 degrees are called complementary angles.

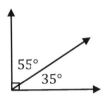

- Two angles with a sum of 180 degrees are Supplementary angles.

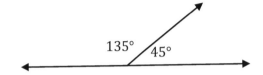

Examples:

Example 1. Find the missing angle.

Solution: Notice that the two angles form a right angle. This means that the angles are complementary, and their sum is 90. 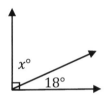 Then: $18 + x = 90 \rightarrow x = 90° - 18° = 72°$

The missing angle is 72 degrees. $x = 72°$

Example 2. Angles Q and S are supplementary. What is the measure of angle Q if angle S is 35 degrees?

Solution: Q and S are supplementary $\rightarrow Q + S = 180 \rightarrow Q + 35 = 180 \rightarrow$
$$Q = 180 - 35 = 145$$

Example 3. Angles x and y are complementary. What is the measure of angle x if angle y is 16 degrees?

Solution: Angles x and y are complementary $\rightarrow x + y = 90 \rightarrow x + 16 = 90 \rightarrow$
$$x = 90 - 16 = 74$$

Parallel lines and Transversals

- When a line (transversal) intersects two parallel lines in the same plane, eight angles are formed. In the following diagram, a transversal intersects two parallel lines. Angles 1, 3, 5, and 7 are congruent. Angles 2, 4, 6, and 8 are also congruent.

- In the following diagram, the following angles are supplementary angles (their sum is 180):

  ❖ Angles 1 and 8

  ❖ Angles 2 and 7

  ❖ Angles 3 and 6

  ❖ Angles 4 and 5

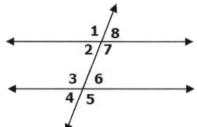

Example:

In the following diagram, two parallel lines are cut by a transversal. What is the value of x?

Solution: The two angles $3x - 15$ and $2x + 7$ are equivalent.

That is: $3x - 15 = 2x + 7$

Now, solve for x:

$3x - 15 + 15 = 2x + 7 + 15$

$\rightarrow 3x = 2x + 22 \rightarrow 3x - 2x = 2x + 22 - 2x \rightarrow$

$x = 22$

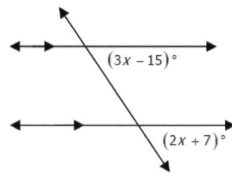

Triangles

- In any triangle, the sum of all angles is 180 degrees.

- Area of a triangle $= \frac{1}{2} (base \times height)$

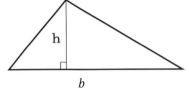

Examples:

Example 1. What is the area of this triangles?

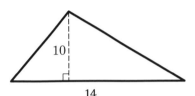

Solution: Use the area formula:
Area $= \frac{1}{2}(base \times height)$
$base = 14$ and $height = 10$, Then:
Area $= \frac{1}{2}(14 \times 10) = \frac{1}{2}(140) = 70$

Example 2. What is the area of this triangles?

Solution: Use the area formula:

Area $= \frac{1}{2}(base \times height)$
$base = 16$ and $height = 8$; Area $= \frac{1}{2}(16 \times 8) = \frac{128}{2} = 64$

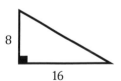

Example 3. What is the missing angle in this triangle?

Solution:

In any triangle, the sum of all angles is 180 degrees.
Let x be the missing angle.
Then: $55 + 80 + x = 180 \rightarrow 135 + x = 180 \rightarrow$
$\qquad x = 180 - 135 = 45$
The missing angle is 45 degrees.

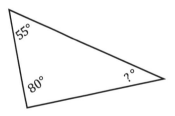

Special Right Triangles

- A special right triangle is a triangle whose sides are in a particular ratio. Two special right triangles are $45° - 45° - 90°$ and $30° - 60° - 90°$ triangles.

- In a special $45° - 45° - 90°$ triangle, the three angles are $45°$, $45°$ and $90°$. The lengths of the sides of this triangle are in the ratio of $1:1:\sqrt{2}$.

- In a special triangle $30° - 60° - 90°$, the three angles are $30° - 60° - 90°$. The lengths of this triangle are in the ratio of $1:\sqrt{3}:2$.

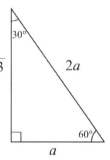

Examples:

Example 1. Find the length of the hypotenuse of a right triangle if the length of the other two sides are both 4 inches.

Solution: this is a right triangle with two equal sides. Therefore, it must be a $45° - 45° - 90°$ triangle. Two equivalent sides are 4 inches. The ratio of sides: $x:x:x\sqrt{2}$

The length of the hypotenuse is $4\sqrt{2}$ inches. $x:x:x\sqrt{2} \rightarrow 4:4:4\sqrt{2}$

Example 2. The length of the hypotenuse of a right triangle is 6 inches. What are the lengths of the other two sides if one angle of the triangle is $30°$?

Solution: The hypotenuse is 6 inches and the triangle is a $30° - 60° - 90°$ triangle. Then, one side of the triangle is 3 (it's half the side of the hypotenuse) and the other side is $3\sqrt{3}$. (it's the smallest side times $\sqrt{3}$)

$x:x\sqrt{3}:2x \rightarrow x = 3 \rightarrow x:x\sqrt{3}:2x = 3:3\sqrt{3}:6$

Polygons

-   The perimeter of a square $= 4 \times side = 4s$

-   The perimeter of a rectangle $= 2(width + length)$

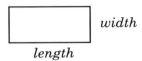

-   The perimeter of trapezoid $= a + b + c + d$

-   The perimeter of a regular hexagon $= 6a$

-   The perimeter of a parallelogram $= 2(l + w)$

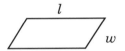

Examples:

Example 1. Find the perimeter of following regular hexagon.

Solution: Since the hexagon is regular, all sides are equal.
Then, the perimeter of the hexagon $= 6 \times (one\ side)$
The perimeter of the hexagon $= 6 \times (one\ side) = 6 \times 8 = 48\ m$

Example 2. Find the perimeter of following trapezoid.

Solution: The perimeter of a trapezoid $= a + b + c + d$
The perimeter of the trapezoid $= 7 + 8 + 8 + 10 = 33\ ft$

Circles

- In a circle, variable r is usually used for the radius and d for diameter.

- *Area of a circle* $= \pi r^2$ (π is about 3.14)

- *Circumference of a circle* $= 2\pi r$

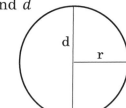

Examples:

Example 1. Find the area of this circle. ($\pi = 3.14$)

Solution:

Use area formula: $Area = \pi r^2$

$r = 6 \, in \rightarrow Area = \pi(6)^2 = 36\pi$, $\pi = 3.14$

Then: $Area = 36 \times 3.14 = 113.04 \, in^2$

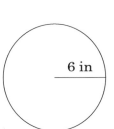

Example 2. Find the Circumference of this circle. ($\pi = 3.14$)

Solution:

Use Circumference formula: $Circumference = 2\pi r$

$r = 8 \, cm \rightarrow Circumference = 2\pi(8) = 16\pi$

$\pi = 3.14$, Then: $Circumference = 16 \times 3.14 = 50.24 \, cm$

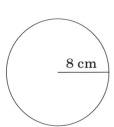

Example 3. Find the area of this circle.

Solution:

Use area formula: $Area = \pi r^2$

$r = 9 \, in$, Then: $Area = \pi(9)^2 = 81\pi$, $\pi = 3.14$

$$Area = 81 \times 3.14 = 254.34 \, in^2$$

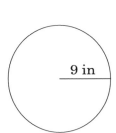

Trapezoids

- A quadrilateral with at least one pair of parallel sides is a trapezoid.

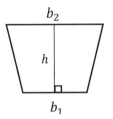

- Area of a trapezoid $= \frac{1}{2}h(b_1 + b_2)$

Examples:

Example 1. Calculate the area of this trapezoid.

Solution:

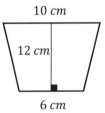

Use area formula: $A = \frac{1}{2}h(b_1 + b_2)$

$b_1 = 6\ cm$, $b_2 = 10\ cm$ and $h = 12\ cm$

Then: $A = \frac{1}{2}(12)(10 + 6) = 6(16) = 96\ cm^2$

Example 2. Calculate the area of this trapezoid.

Solution:

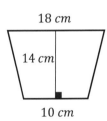

Use area formula: $A = \frac{1}{2}h(b_1 + b_2)$

$b_1 = 10\ cm$, $b_2 = 18\ cm$ and $h = 14\ cm$

Then: $A = \frac{1}{2}(14)(10 + 18) = 196\ cm^2$

Cubes

- A cube is a three-dimensional solid object bounded by six square sides.

- Volume is the measure of the amount of space inside of a solid figure, like a cube, ball, cylinder or pyramid.

- The volume of a cube $= (one\ side)^3$

- The surface area of a cube $= 6 \times (one\ side)^2$

Examples:

Example 1. Find the volume and surface area of this cube.

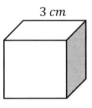

Solution: Use volume formula: $volume = (one\ side)^3$

Then: $volume = (one\ side)^3 = (3)^3 = 27\ cm^3$

Use surface area formula:

$surface\ area\ of\ a\ cube$: $6(one\ side)^2 = 6(3)^2 = 6(9) = 54\ cm^2$

Example 2. Find the volume and surface area of this cube.

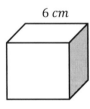

Solution: Use volume formula: $volume = (one\ side)^3$

Then: $volume = (one\ side)^3 = (6)^3 = 216\ cm^3$

Use surface area formula:

$surface\ area\ of\ a\ cube$: $6(one\ side)^2 = 6(6)^2 = 6(36) = 216\ cm^2$

Example 3. Find the volume and surface area of this cube.

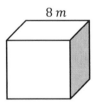

Solution: Use volume formula: $volume = (one\ side)^3$

Then: $volume = (one\ side)^3 = (8)^3 = 512\ m^3$

Use surface area formula:

$surface\ area\ of\ a\ cube$: $6(one\ side)^2 = 6(8)^2 = 6(64) = 384\ m^2$

Rectangular Prisms

- A rectangular prism is a solid 3-dimensional object with six rectangular faces.

- The volume of a Rectangular prism = *Length × Width × Height*

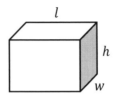

$Volume = l \times w \times h$

$Surface\ area = 2 \times (wh + lw + lh)$

Examples:

Example 1. Find the volume and surface area of this rectangular prism.

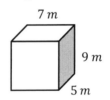

Solution: Use volume formula: $Volume = l \times w \times h$

Then: $Volume = 7 \times 5 \times 9 = 315\ m^3$

Use surface area formula: $Surface\ area = 2 \times (wh + lw + lh)$

Then: $Surface\ area = 2 \times ((5 \times 9) + (7 \times 5) + (7 \times 9))$

$\qquad\qquad = 2 \times (45 + 35 + 63) = 2 \times (143) = 286\ m^2$

Example 2. Find the volume and surface area of this rectangular prism.

Solution: Use volume formula: $Volume = l \times w \times h$

Then: $Volume = 9 \times 6 \times 12 = 648\ m^3$

Use surface area formula: $Surface\ area = 2 \times (wh + lw + lh)$

Then: $Surface\ area = 2 \times ((6 \times 12) + (9 \times 6) + (9 \times 12))$

$\qquad\qquad = 2 \times (72 + 54 + 108) = 2 \times (234) = 468\ m^2$

Cylinder

- A cylinder is a solid geometric figure with straight parallel sides and a circular or oval cross-section.

- *Volume of a Cylinder* $= \pi (radius)^2 \times height$, $\pi \approx 3.14$

- *Surface area of a cylinder* $= 2\pi r^2 + 2\pi rh$

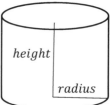

Examples:

Example 1. Find the volume and Surface area of the follow Cylinder.

Solution: Use volume formula:

$Volume = \pi (radius)^2 \times height$

Then: $Volume = \pi(4)^2 \times 10 = 16\pi \times 10 = 160\pi$

$\pi = 3.14$, then: $Volume = 160\pi = 160 \times 3.14 = 502.4 \ cm^3$

Use surface area formula: $Surface\ area = 2\pi r^2 + 2\pi rh$

Then: $2\pi(4)^2 + 2\pi(4)(10) = 2\pi(16) + 2\pi(40) = 32\pi + 80\pi = 112\pi$

$\pi = 3.14$, Then: $Surface\ area = 112 \times 3.14 = 351.68 \ cm^2$

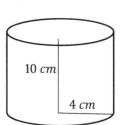

Example 2. Find the volume and Surface area of the follow Cylinder.

Solution: Use volume formula:

$Volume = \pi (radius)^2 \times height$

Then: $Volume = \pi(5)^2 \times 8 = 25\ \pi \times 8 = 200\pi$

$\pi = 3.14$, Then: $Volume = 200\pi = 628 \ cm^3$

Use surface area formula: $Surface\ area = 2\pi r^2 + 2\pi rh$

Then: $= 2\pi(5)^2 + 2\pi(5)(8) = 2\pi(25) + 2\pi(40) = 50\pi + 80\pi = 130\pi$

$\pi = 3.14$ then: $Surface\ area = 130 \times 3.14 = 408.2 \ cm^2$

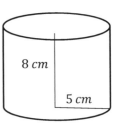

Chapter 11: Practices

✏ Find the missing side?

1)
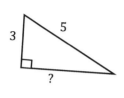

2)

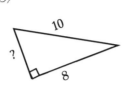

3)
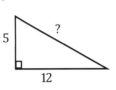

4)

✏ Find the measure of the unknown angle in each triangle.

5)
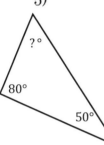

6)
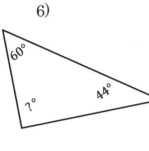

7)
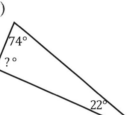

8)
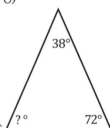

✏ Find the area of each triangle.

9)
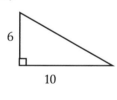

10)

11)
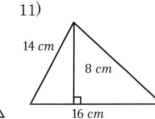

12)
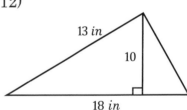

✏ Find the perimeter or circumference of each shape.

13)
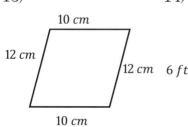

14)
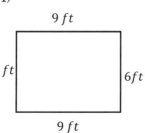

15)
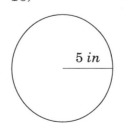

16) *regular hexagon*
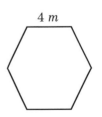

✎ Find the area of each trapezoid.

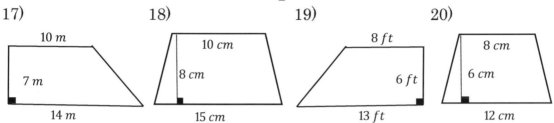

17)   18)   19)   20)

10 m / 7 m / 14 m

10 cm / 8 cm / 15 cm

8 ft / 6 ft / 13 ft

8 cm / 6 cm / 12 cm

✎ Find the volume of each cube.

21)   22)   23)   24)

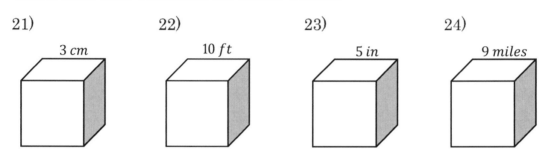

3 cm   10 ft   5 in   9 miles

✎ Find the volume of each Rectangular Prism.

25)   26)   27)

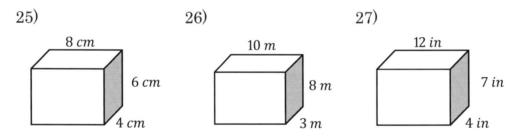

8 cm / 6 cm / 4 cm

10 m / 8 m / 3 m

12 in / 7 in / 4 in

✎ Find the volume of each Cylinder. Round your answer to the nearest tenth. (π = 3.14)

28)   29)   30)

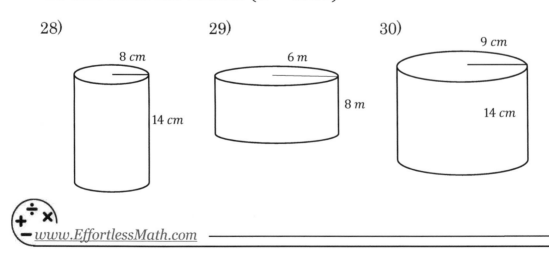

8 cm / 14 cm

6 m / 8 m

9 cm / 14 cm

Chapter 11: Answers

1) 4

2) 15

3) 6

4) 13

5) 50

6) 76

7) 84

8) 70

9) 30

10) 49.5

11) $64 \ cm^2$

12) $90 \ in^2$

13) $44 \ cm$

14) $30 \ ft$

15) $10 \ \pi \approx 31.4 \ in$

16) $24 \ m$

17) $84 \ m^2$

18) $100 \ cm^2$

19) $63 \ ft^2$

20) $60 \ cm^2$

21) $27 \ cm^3$

22) $1,000 \ ft^3$

23) $125 \ in^3$

24) $729 \ mi^3$

25) $192 \ cm^3$

26) $240 \ m^3$

27) $336 \ in^3$

28) $2,813.44 \ cm^3$

29) $904.32 \ m^3$

30) $3,560.76 \ cm^3$

CHAPTER

12 Statistics

Math topics that you'll learn in this chapter:

- ☑ Mean, Median, Mode, and Range of the Given Data
- ☑ Pie Graph
- ☑ Probability Problems
- ☑ Permutations and Combinations

115

Mean, Median, Mode, and Range of the Given Data

- **Mean:** $\dfrac{sum\ of\ the\ data}{total\ number\ of\ data\ entires}$

- **Mode:** the value in the list that appears most often

- **Median:** is the middle number of a group of numbers arranged in order by size.

- **Range:** the difference of the largest value and smallest value in the list

Examples:

Example 1. What is the mode of these numbers? $5, 6, 8, 6, 8, 5, 3, 5$

Solution: Mode: the value in the list that appears most often.
Therefore, the mode is number 5. There are three number 5 in the data.

Example 2. What is the median of these numbers? $6, 11, 15, 10, 17, 20, 7$

Solution: Write the numbers in order: $6, 7, 10, 11, 15, 17, 20$
The median is the number in the middle. Therefore, the median is 11.

Example 3. What is the mean of these numbers? $7, 2, 3, 2, 4, 8, 7, 5$

Solution: Mean: $\dfrac{sum\ of\ the\ data}{total\ number\ of\ data\ entires} = \dfrac{7+2+3+2+4+8+7+5}{8} = \dfrac{38}{8} = 4.75$

Example 4. What is the range in this list? $3, 7, 12, 6, 15, 20, 8$

Solution: Range is the difference of the largest value and smallest value in the list. The largest value is 20 and the smallest value is 3.
Then: $20 - 3 = 17$

Pie Graph

- A Pie Graph (Pie Chart) is a circle chart divided into sectors, each sector represents the relative size of each value.

- Pie charts represent a snapshot of how a group is broken down into smaller pieces.

Example:

A library has 750 books that include Mathematics, Physics, Chemistry, English and History. Use the following graph to answer the questions.

Example 1. What is the number of Mathematics books?

Solution: Number of total books = 750

Percent of Mathematics books = 28%

Then, the number of Mathematics books: $28\% \times 750 = 0.28 \times 750 = 210$

Example 2. What is the number of History books?

Solution: Number of total books = 750

Percent of History books = 12%

Then: $0.12 \times 750 = 90$

Example 3. What is the number of Chemistry books in the library?

Solution: Number of total books = 750

Percent of Chemistry books = 22%

Then: $0.22 \times 750 = 165$

Probability Problems

- Probability is the likelihood of something happening in the future. It is expressed as a number between zero (can never happen) to 1 (will always happen).

- Probability can be expressed as a fraction, a decimal, or a percent.

- Probability formula: $Probability = \frac{number\ of\ desired\ outcomes}{number\ of\ total\ outcomes}$

Examples:

Example 1. Anita's trick–or–treat bag contains 10 pieces of chocolate, 16 suckers, 16 pieces of gum, 22 pieces of licorice. If she randomly pulls a piece of candy from her bag, what is the probability of her pulling out a piece of sucker?

Solution: Probability $= \frac{number\ of\ desired\ outcomes}{number\ of\ total\ outcomes}$

Probability of pulling out a piece of sucker $= \dfrac{16}{10 + 16 + 16 + 22} = \dfrac{16}{64} = \dfrac{1}{4}$

Example 2. A bag contains 20 balls: four green, five black, eight blue, a brown, a red and one white. If 19 balls are removed from the bag at random, what is the probability that a brown ball has been removed?

Solution: If 19 balls are removed from the bag at random, there will be one ball in the bag. The probability of choosing a brown ball is 1 out of 20. Therefore, the probability of not choosing a brown ball is 19 out of 20 and the probability of having not a brown ball after removing 19 balls is the same. The answer is: $\frac{19}{20}$

Permutations and Combinations

Factorials are products, indicated by an exclamation mark. For example,
$4! = 4 \times 3 \times 2 \times 1$ (Remember that $0!$ is defined to be equal to 1)

- **Permutations:** The number of ways to choose a sample of k elements from a set of n distinct objects where order does matter, and replacements are not allowed. For a permutation problem, use this formula:

$$_nP_k = \frac{n!}{(n-k)!}$$

- **Combination:** The number of ways to choose a sample of r elements from a set of n distinct objects where order does not matter, and replacements are not allowed. For a combination problem, use this formula:

$$_nC_r = \frac{n!}{r!\,(n-r)!}$$

Examples:

Example 1. How many ways can the first and second place be awarded to 7 people?

Solution: Since the order matters, (the first and second place are different!) we need to use permutation formula where n is 7 and k is 2. Then: $\frac{n!}{(n-k)!} = \frac{7!}{(7-2)!} = \frac{7!}{5!} = \frac{7 \times 6 \times 5!}{5!}$, remove $5!$ from both sides of the fraction. Then: $\frac{7 \times 6 \times 5!}{5!} = 7 \times 6 = 42$

Example 2. How many ways can we pick a team of 3 people from a group of 8?

Solution: Since the order doesn't matter, we need to use a combination formula where n is 8 and r is 3.
Then: $\frac{n!}{r!\,(n-r)!} = \frac{8!}{3!\,(8-3)!} = \frac{8!}{3!\,(5)!} = \frac{8 \times 7 \times 6 \times 5!}{3!\,(5)!} = \frac{8 \times 7 \times 6}{3 \times 2 \times 1} = \frac{336}{6} = 56$

Chapter 12: Practices

✍ Find the values of the Given Data.

1) 6, 11, 5, 3, 6

Mode: _____   Range: _____

Mean: _____   Median: _____

2) 4, 9, 1, 9, 6, 7

Mode: _____   Range: _____

Mean: _____   Median: _____

3) 10, 3, 6, 10, 4, 15

Mode: _____   Range: _____

Mean: _____   Median: _____

4) 12, 4, 8, 9, 3, 12, 15

Mode: _____   Range: _____

Mean: _____   Median: _____

✍ The circle graph below shows all Bob's expenses for last month. Bob spent $790 on his Rent last month.

5) How much did Bob's total expenses last month? _____

6) How much did Bob spend for foods last month? _____

7) How much did Bob spend for his bills last month?

_____

8) How much did Bob spend on his car last month? _____

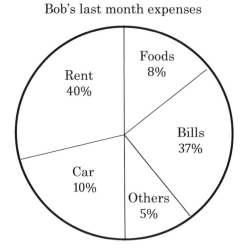

Bob's last month expenses

Rent 40%

Foods 8%

Bills 37%

Car 10%

Others 5%

✎ Solve.

9) Bag A contains 8 red marbles and 6 green marbles. Bag B contains 5 black marbles and 7 orange marbles. What is the probability of selecting a green marble at random from bag A? What is the probability of selecting a black marble at random from Bag B?

_____   _____

✎ Solve.

10) Susan is baking cookies. She uses sugar, flour, butter, and eggs. How many different orders of ingredients can she try? _____

11) Jason is planning for his vacation. He wants to go to museum, go to the beach, and play volleyball. How many different ways of ordering are there for him? _____

12) In how many ways can a team of 6 basketball players choose a captain and co-captain? _____

13) How many ways can you give 5 balls to your 8 friends? _____

14) A professor is going to arrange her 5 students in a straight line. In how many ways can she do this? _____

15) In how many ways can a teacher chooses 12 out of 15 students?

_____

Chapter 12: Answers

1) Mode: 6, Range: 8, Mean: 6.2, Median: 6

2) Mode: 9, Range:8, Mean: 6, Median: 6.5

3) Mode: 10, Range: 12, Mean: 8, Median: 8

4) Mode: 12, Range: 12, Mean: 9, Median: 9

5) $1,975

6) $158

7) $730.75

8) $197.50

9) $\frac{3}{7}, \frac{5}{12}$

10) 24

11) 6

12) 30 (it's a permutation problem)

13) 56 (it's a combination problem)

14) 120

15) 455 (it's a combination problem)

CHAPTER

13 Functions Operations

Math topics that you'll learn in this chapter:

- ☑ Function Notation and Evaluation
- ☑ Adding and Subtracting Functions
- ☑ Multiplying and Dividing Functions
- ☑ Composition of Functions

123

Function Notation and Evaluation

- Functions are mathematical operations that assign unique outputs to given inputs.

- Function notation is the way a function is written. It is meant to be a precise way of giving information about the function without a rather lengthy written explanation.

- The most popular function notation is $f(x)$ which is read "f of x". Any letter can name a function. for example: $g(x)$, $h(x)$, etc.

- To evaluate a function, plug in the input (the given value or expression) for the function's variable (place holder, x).

Examples:

Example 1. Evaluate: $f(x) = x + 6$, find $f(2)$

Solution: Substitute x with 2:
Then: $f(x) = x + 6 \rightarrow f(2) = 2 + 6 \rightarrow f(2) = 8$

Example 2. Evaluate: $w(x) = 3x - 1$, find $w(4)$.

Solution: Substitute x with 4:
Then: $w(x) = 3x - 1 \rightarrow w(4) = 3(4) - 1 = 12 - 1 = 11$

Example 3. Evaluate: $f(x) = 2x^2 + 4$, find $f(-1)$.

Solution: Substitute x with -1:
Then: $f(x) = 2x^2 + 4 \rightarrow f(-1) = 2(-1)^2 + 4 \rightarrow f(-1) = 2 + 4 = 6$

Example 4. Evaluate: $h(x) = 4x^2 - 9$, find $h(2a)$.

Solution: Substitute x with $2a$:
Then: $h(x) = 4x^2 - 9 \rightarrow h(2a) = 4(2a)^2 - 9 \rightarrow h(2a) = 4(4a^2) - 9 = 16a^2 - 9$

Adding and Subtracting Functions

- Just like we can add and subtract numbers and expressions, we can add or subtract functions and simplify or evaluate them. The result is a new function.

- For two functions $f(x)$ and $g(x)$, we can create two new functions:

$$(f + g)(x) = f(x) + g(x) \text{ and } (f\text{-}g)(x) = f(x) - g(x)$$

Examples:

Example 1. $g(x) = 2x - 2$, $f(x) = x + 1$, Find: $(g + f)(x)$

Solution: $(g + f)(x) = g(x) + f(x)$
Then: $(g + f)(x) = (2x - 2) + (x + 1) = 2x - 2 + x + 1 = 3x - 1$

Example 2. $f(x) = 4x - 3$, $g(x) = 2x - 4$, Find: $(f - g)(x)$

Solution: $(f - g)(x) = f(x) - g(x)$
Then: $(f - g)(x) = (4x - 3) - (2x - 4) = 4x - 3 - 2x + 4 = 2x + 1$

Example 3. $g(x) = x^2 + 2$, $f(x) = x + 5$, Find: $(g + f)(x)$

Solution: $(g + f)(x) = g(x) + f(x)$
Then: $(g + f)(x) = (x^2 + 2) + (x + 5) = x^2 + x + 7$

Example 4. $f(x) = 5x^2 - 3$, $g(x) = 3x + 6$, Find: $(f - g)(3)$

Solution: $(f - g)(x) = f(x) - g(x)$
Then: $(f - g)(x) = (5x^2 - 3) - (3x + 6) = 5x^2 - 3 - 3x - 6 = 5x^2 - 3x - 9$
Substitute x with 3: $(f - g)(3) = 5(3)^2 - 3(3) - 9 = 45 - 9 - 9 = 27$

Multiplying and Dividing Functions

- Just like we can multiply and divide numbers and expressions, we can multiply and divide two functions and simplify or evaluate them.

- For two functions $f(x)$ and $g(x)$, we can create two new functions:

$$(f.g)(x) = f(x).g(x) \text{ and } \left(\frac{f}{g}\right)(x) = \frac{f(x)}{g(x)}$$

Examples:

Example 1. $g(x) = x + 3$, $f(x) = x + 4$, Find: $(g.f)(x)$

Solution:

$(g.f)(x) = g(x).f(x) = (x+3)(x+4) = x^2 + 4x + 3x + 12 = x^2 + 7x + 12$

Example 2. $f(x) = x + 6$, $h(x) = x - 9$, Find: $\left(\frac{f}{h}\right)(x)$

Solution: $\left(\frac{f}{h}\right)(x) = \frac{f(x)}{h(x)} = \frac{x+6}{x-9}$

Example 3. $g(x) = x + 7$, $f(x) = x - 3$, Find: $(g.f)(2)$

Solution: $(g.f)(x) = g(x).f(x) = (x+7)(x-3) = x^2 - 3x + 7x - 21$

$$g(x).f(x) = x^2 + 4x - 21$$

Substitute x with 2: $(g.f)(x) = (2)^2 + 4(2) - 21 = 4 + 8 - 21 = -9$

Example 4. $f(x) = x + 3$, $h(x) = 2x - 4$, Find: $\left(\frac{f}{h}\right)(3)$

Solution: $\left(\frac{f}{h}\right)(x) = \frac{f(x)}{h(x)} = \frac{x+3}{2x-4}$

Substitute x with 3: $\left(\frac{f}{h}\right)(x) = \frac{x+3}{2x-4} = \frac{3+3}{2(3)-4} = \frac{6}{2} = 3$

Composition of Functions

- "Composition of functions" simply means combining two or more functions in a way where the output from one function becomes the input for the next function.

- The notation used for composition is: $(fog)(x) = f(g(x))$ and is read

    "f composed with g of x" or "f of g of x".

Examples:

Example 1. Using $f(x) = 2x + 3$ and $g(x) = 5x$, find: $(fog)(x)$

Solution: $(fog)(x) = f(g(x))$. Then: $(fog)(x) = f(g(x)) = f(5x)$
Now find $f(5x)$ by substituting x with $5x$ in $f(x)$ function.
Then: $f(x) = 2x + 3$; $(x \rightarrow 5x) \rightarrow f(5x) = 2(5x) + 3 = 10x + 3$

Example 2. Using $f(x) = 3x - 1$ and $g(x) = 2x - 2$, find: $(gof)(5)$

Solution: $(fog)(x) = f(g(x))$. Then: $(gof)(x) = g(f(x)) = g(3x - 1)$,
Now substitute x in $g(x)$ *by* $(3x - 1)$.
Then: $g(3x - 1) = 2(3x - 1) - 2 = 6x - 2 - 2 = 6x - 4$
Substitute x with 5: $(gof)(5) = g(f(x)) = 6x - 4 = 6(5) - 4 = 26$

Example 3. Using $f(x) = 2x^2 - 5$ and $g(x) = x + 3$, find: $f(g(3))$

Solution: First, find $g(3)$: $g(x) = x + 3 \rightarrow g(3) = 3 + 3 = 6$
Then: $f(g(3)) = f(6)$. Now, find $f(6)$ by substituting x with 6 in $f(x)$ function.
$f(g(3)) = f(6) = 2(6)^2 - 5 = 2(36) - 5 = 67$

Chapter 13: Practices

✍ Evaluate each function.

1) $g(n) = 2n + 5$, find $g(2)$

_____

2) $h(x) = 5n - 9$, find $h(4)$

_____

3) $k(n) = 10 - 6n$, find $k(2)$

_____

4) $g(x) = -5x + 6$, find $g(-2)$

_____

5) $k(n) = -8n + 3$, find $k(-6)$

_____

6) $w(n) = -2n - 9$, find $w(-5)$

_____

✍ Perform the indicated operation.

7) $f(x) = x + 6$
   $g(x) = 3x + 2$
   Find $(f - g)(x)$

_____

8) $g(x) = x - 9$
   $f(x) = 2x - 1$
   Find $(g - f)(x)$

_____

9) $h(t) = 5t + 6$
   $g(t) = 2t + 4$
   Find $(h + g)(x)$

_____

10) $g(a) = -6a + 1$
    $f(a) = 3a^2 - 3$
    Find $(g + f)(5)$

_____

11) $g(x) = 7x - 1$
    $h(x) = -4x^2 + 2$
    Find $(g - h)(-3)$

_____

12) $h(x) = -x^2 - 1$
    $g(x) = -7x - 1$
    Find $(h - g)(-5)$

_____

✍ Perform the indicated operation.

13) $g(x) = x + 3$

$f(x) = x + 1$

Find $(g.f)(x)$

_____

14) $f(x) = 4x$

$h(x) = x - 6$

Find $(f.h)(x)$

_____

15) $g(a) = a - 8$

$h(a) = 4a - 2$

Find $(g.h)(3)$

_____

16) $f(x) = 6x + 2$

$h(x) = 5x - 1$

Find $\left(\frac{f}{h}\right)(-2)$

_____

17) $f(x) = 7a - 1$

$g(x) = -5 - 2a$

Find $\left(\frac{f}{g}\right)(-4)$

_____

18) $g(a) = a^2 - 4$

$f(a) = a + 6$

Find $\left(\frac{g}{f}\right)(-3)$

_____

✍ Using $f(x) = 4x + 3$ and $g(x) = x - 7$, find:

19) $g\big(f(2)\big) = $ _____

20) $g\big(f(-2)\big) = $ _____

21) $f\big(g(4)\big) = $ _____

22) $f\big(f(7)\big) = $ _____

23) $g\big(f(5)\big) = $ _____

24) $g\big(f(-5)\big) = $ _____

25) $g\big(f(7)\big) = $ _____

26) $g\big(f(-3)\big) = $ _____

27) $f\big(g(-6)\big) = $ _____

Chapter 13: Answers

1) 9

2) 11

3) -2

4) 16

5) 51

6) 1

7) $-2x - 4$

8) $-x - 8$

9) $7t + 10$

10) 43

11) 12

12) -60

13) $x^2 + 4x + 3$

14) $4x^2 - 24x$

15) -50

16) $\frac{10}{11}$

17) $-\frac{29}{3}$

18) $\frac{5}{3}$

19) 4

20) -12

21) -9

22) 127

23) 16

24) -24

25) 24

26) -16

27) -49

Time to Test

Time to refine your skill with a practice examination

Take a practice SSAT Upper Level Mathematics Test to simulate the test day experience. After you've finished, score your test using the answer keys.

Before You Start

- You'll need a pencil and a timer to take the test.

- Each test contains 25 multiple-choice questions. For each question, there are five possible answers. Choose which one is best.

- After you've finished the test, review the answer key to see where you went wrong.

- Use the answer sheet provided to record your answers. (You can cut it out or photocopy it)

- You will receive 1 point for every correct answer, and you will lose $\frac{1}{4}$ point for each incorrect answer. There is no penalty for skipping a question.

Calculators are NOT permitted for the SSAT Upper Level Test

Good Luck!

SSAT Upper Level Math Practice Test 1

2021

Two Parts

Total number of questions: 50

Section 1: 25 questions

Section 2: 25 questions

Total time for two parts: 60 Minutes

SSAT Upper Level Math Practice Test 1 Answer Sheet

Remove (or photocopy) this answer sheet and use it to complete the practice test.

SSAT Upper Level Mathematics Practice Test 1 Answer Sheet

SSAT Upper Level Practice Section 1

1	(A) (B) (C) (D) (E)	11 (A) (B) (C) (D) (E)	21 (A) (B) (C) (D) (E)
2	(A) (B) (C) (D) (E)	12 (A) (B) (C) (D) (E)	22 (A) (B) (C) (D) (E)
3	(A) (B) (C) (D) (E)	13 (A) (B) (C) (D) (E)	23 (A) (B) (C) (D) (E)
4	(A) (B) (C) (D) (E)	14 (A) (B) (C) (D) (E)	24 (A) (B) (C) (D) (E)
5	(A) (B) (C) (D) (E)	15 (A) (B) (C) (D) (E)	25 (A) (B) (C) (D) (E)
6	(A) (B) (C) (D) (E)	16 (A) (B) (C) (D) (E)	
7	(A) (B) (C) (D) (E)	17 (A) (B) (C) (D) (E)	
8	(A) (B) (C) (D) (E)	18 (A) (B) (C) (D) (E)	
9	(A) (B) (C) (D) (E)	19 (A) (B) (C) (D) (E)	
10	(A) (B) (C) (D) (E)	20 (A) (B) (C) (D) (E)	

SSAT Upper Level Practice Section 2

1	(A) (B) (C) (D) (E)	11 (A) (B) (C) (D) (E)	21 (A) (B) (C) (D) (E)
2	(A) (B) (C) (D) (E)	12 (A) (B) (C) (D) (E)	22 (A) (B) (C) (D) (E)
3	(A) (B) (C) (D) (E)	13 (A) (B) (C) (D) (E)	23 (A) (B) (C) (D) (E)
4	(A) (B) (C) (D) (E)	14 (A) (B) (C) (D) (E)	24 (A) (B) (C) (D) (E)
5	(A) (B) (C) (D) (E)	15 (A) (B) (C) (D) (E)	25 (A) (B) (C) (D) (E)
6	(A) (B) (C) (D) (E)	16 (A) (B) (C) (D) (E)	
7	(A) (B) (C) (D) (E)	17 (A) (B) (C) (D) (E)	
8	(A) (B) (C) (D) (E)	18 (A) (B) (C) (D) (E)	
9	(A) (B) (C) (D) (E)	19 (A) (B) (C) (D) (E)	
10	(A) (B) (C) (D) (E)	20 (A) (B) (C) (D) (E)	

SSAT Upper Level Math
Practice Test 1

Section 1

25 questions

Total time for this section: 30 Minutes

You may NOT use a calculator for this test.

1) What is the value of the "9" in number 131.493?

A. 9 ones

B. 9 tenths

C. 9 hundredths

D. 9 tens

E. 9 thousandths

2) If $x - 20 = -20$, then $x \times 20 = $?

A. 0

B. 10

C. 20

D. 40

E. 60

3) $0.04 \times 13.00 = ?$

A. 5.2

B. 52.00

C. 0.52

D. 5.02

E. 0.052

4) If Logan ran 3.5 miles in half an hour, his average speed was?

A. 2.25 *miles per hour*

B. 3.5 *miles per hour*

C. 3.75 *miles per hour*

D. 4.26 *miles per hour*

E. 7 *miles per hour*

5) Given the diagram, what is the perimeter of the quadrilateral?

A. 54

B. 70

C. 720

D. 26740

E. 55480

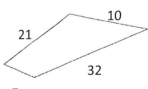

6) A pizza maker has M pounds of flour to make pizzas. After he has used 75 pounds of flour, how much flour is left? The expression that correctly represents the quantity of flour left is:

A. $75 + M$

B. $\frac{75}{M}$

C. $75 - M$

D. $M - 75$

E. $75M$

7) Mia plans to buy a bracelet for every one of her 18 friends for their party. There are four bracelets in each pack. How many packs must she buy?

A. 2

B. 3

C. 4

D. 5

E. 10

8) The distance between cities A and B is approximately 3,600 miles. If Alice drives an average of 70 miles per hour, how many hours will it take Alice to drive from city A to city B?

A. *Approximately* 70 *hours*

B. *Approximately* 68 *hours*

C. *Approximately* 51*hours*

D. *Approximately* 37 *hours*

E. *Approximately* 21 *hours*

9) In a classroom of 60 students, 30 are female. What percentage of the class is male?

A. 54%

B. 50%

C. 30%

D. 26%

E. 10%

10) A steak dinner at a restaurant costs $9.20. If a man buys a steak dinner for himself and 4 friends, what will the total cost be?

A. $33

B. $46

C. $28

D. $22.5

E. 12

11) An employee's rating on performance appraisals for the last three quarters were 93, 40 and 87. If the required yearly average to qualify for the promotion is 97, what rating should the fourth quarter be?

A. 158

B. 168

C. 178

D. 188

E. 198

12) Two third of 15 is equal to $\frac{2}{5}$ of what number?

A. 15

B. 25

C. 35

D. 45

E. 55

13) A cruise line ship left Port A and traveled 60 miles due west and then 80 miles due north. At this point, what is the shortest distance from the cruise to port A?

A. 100 *miles*

B. 120 *miles*

C. 130 *miles*

D. 150 *miles*

E. 170 *miles*

14) Last week 25,000 fans attended a football match. This week three times as many bought tickets, but one sixth of them cancelled their tickets. How many are attending this week?

A. 49,000

B. 52,200

C. 62,500

D. 72,500

E. 84,700

15) What is the slope of the line that is perpendicular to the line with equation $8x + y = 14$?

A. $\frac{1}{8}$

B. $-\frac{1}{8}$

C. $\frac{8}{12}$

D. 8

E. -8

16) In 1989, the average worker's income increased $3,000 per year starting from $34,000 annual salary. Which equation represents income greater than average? (I = income, x = number of years after 1989)

A. $I > 3,000\, x + 34,000$

B. $I > -\,3,000\, x + 34,000$

C. $I < -3,000\, x + 34,000$

D. $I < 3,000\, x - 34,000$

E. $I < 34,000\, x + 34,000$

17) In the following figure, MN is 50 cm. How long is ON?

A. 35 cm

B. 30 cm

C. 25 cm

D. 15 cm

E. 5 cm

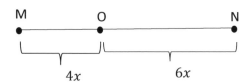

Questions 18 to 20 are based on the following data

The result of a research shows the number of men and women in four cities of a country.

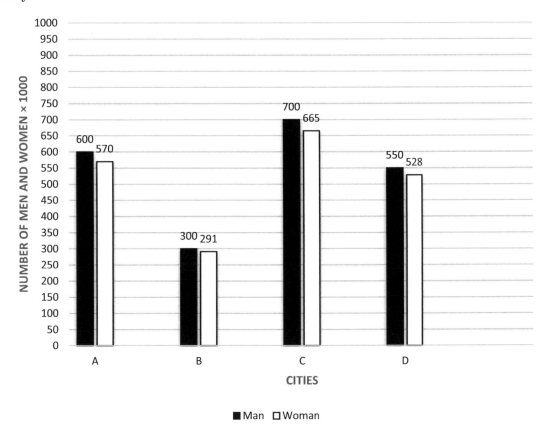

18) How many women should be added to city D until the ratio of women to men will be 1.2?

A. 120

B. 128

C. 132

D. 160

E. 165

19) What's the maximum ratio of woman to man in the four cities?

A. 0.98

B. 0.97

C. 0.96

D. 0.95

E. 0.93

20) What's the ratio of percentage of men in city A to percentage of women in city C?

A. 0.9

B. 0.95

C. 1

D. 1.05

E. 1.5

21) What is the difference of smallest 5-digit number and biggest 5-digit number?

A. 66,666

B. 6,7899

C. 8,8888

D. 8,9999

E. 9,9999

22) Solve the following equation for y?

$$\frac{x}{3+4} = \frac{y}{11-8}$$

A. $\frac{3}{5}x$

B. $\frac{3}{7}x$

C. $3x$

D. x

E. $\frac{1}{7}x$

23) $\sqrt[5]{x^{21}} = ?$

A. $x^4\sqrt[5]{x}$

B. $20x$

C. x^{11}

D. x^{80}

E. x^4

24) Rectangle A has a length of 10 cm and a width of 6 cm, and rectangle B has a length of 6 cm and a width of 4 cm, what is the percent of ratio of the perimeter of rectangle B to rectangle A?

A. 10%

B. 20%

C. 62%

D. 75%

E. 143%

25) John traveled 160 km in 4 hours and Alice traveled 190 km in 5 hours. What is the ratio of the average speed of John to average speed of Alice?

A. 2 : 3

B. 3 : 2

C. 9 : 5

D. 6 : 5

E. 20 : 19

IF YOU FINISH BEFORE TIME IS CALLED, YOU MAY CHECK YOUR WORK ON THIS SECTION ONLY. DO NOT TURN TO ANY OTHER SECTION IN THE TEST.   **STOP**

SSAT Upper Level Math
Practice Test 1

Section 2

25 questions

Total time for this section: 30 Minutes

You may NOT use a calculator for this test.

1) $0.44 \times 12.8 = ?$

A. 4.632

B. 4.965

C. 5.632

D. 5.891

E. 6.695

2) $4\frac{3}{8} \times 5\frac{1}{9} = ?$

A. $22\frac{1}{36}$

B. $22\frac{13}{36}$

C. $23\frac{4}{36}$

D. $23\frac{13}{36}$

E. 23

3) Which of the following is a whole number?

A. $\frac{3}{2} \times \frac{5}{9}$

B. $\frac{1}{3} + \frac{1}{4}$

C. $\frac{40}{6}$

D. $3.5 + 2$

E. $3.5 + \frac{4}{8}$

4) 8 cubed is the same as:

A. 8×8

B. $8 \times 8 \times 8 \times 8$

C. 64

D. 512

E. 15,807

5) If $\frac{2}{3}$ of a number equal to 16 then $\frac{5}{3}$ of the same number is:

A. 40

B. 35

C. 20

D. 14

E. 4

6) A swimming pool holds 3,000 cubic feet of water. The swimming pool is 30 feet long and 10 feet wide. How deep is the swimming pool?

A. 3 feet

B. 5 feet

C. 7 feet

D. 10 feet

E. 12 feet

7) We can put 25 colored pencils in each box and we have 450 colored pencils. How many boxes do we need?

A. 12

B. 15

C. 17

D. 18

E. 19

8) In the figure below, line A is parallel to line B. What is the value of angle x?

A. 45 degree

B. 55 degree

C. 80 degree

D. 120 degree

E. 145 degree

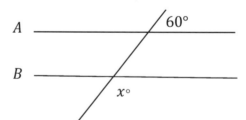

9) A bank is offering 4.5% simple interest on a savings account. If you deposit $13,500, how much interest will you earn in two years?

A. $420

B. $750

C. $6,300

D. $9,400

E. $1,215

10) Sophia purchased a sofa for $504. The sofa is regularly priced at $600. What was the percent discount Sophia received on the sofa?

A. 10%

B. 12%

C. 16%

D. 25%

E. 30%

11) How long does a 527-miles trip take moving at 62 miles per hour (*mph*)?

A. 7 *hours*

B. 7 *hours and* 24 *minutes*

C. 8 *hours and* 24 *minutes*

D. 8 *hours and* 30 *minutes*

E. 9 *hours and* 30 *minutes*

12) A construction company is building a wall. The company can build 40 *cm* of the wall per minute. After 50 minutes construction, $\frac{2}{3}$ of the wall is completed. How high is the wall?

A. 10 *m*

B. 15 *m*

C. 30 *m*

D. 35 *m*

E. 42 *m*

13) When a number is subtracted from 28 and the difference is divided by that number, the result is 3. What is the value of the number?

A. 2

B. 5

C. 7

D. 12

E. 28

14) If car A drives 700 miles in 7 hours and car B drives the same distance in 5 hours, how many miles per hour does car B drive faster than car A?

A. 88 miles per hour

B. 65 miles per hour

C. 40 miles per hour

D. 12 miles per hour

E. 5 miles per hour

15) A company pays its employee $8,000 plus 4% of all sales profit. If x is the number of all sales profit, which of the following represents the employee's revenue?

A. $0.04x$

B. $0.88x - 8,000$

C. $0.04x + 8,000$

D. $0.88x + 8,000$

E. $0.88x$

16) Which of the following shows the numbers in decreasing order?

A. $\frac{1}{5}, \frac{5}{3}, \frac{8}{11}, \frac{2}{3}$

B. $\frac{5}{3}, \frac{8}{11}, \frac{2}{3}, \frac{1}{5}$

C. $\frac{8}{11}, \frac{2}{3}, \frac{5}{3}, \frac{1}{5}$

D. $\frac{2}{3}, \frac{8}{11}, \frac{5}{3}, \frac{1}{5}$

E. None of the above

17) The ratio of boys to girls in a school is $3:2$. If there are 700 students in a school, how many boys are in the school?

A. 550

B. 500

C. 460

D. 420

E. 280

18) $\frac{(8+6)^2}{2} + 6 = ?$

A. 51

B. 80

C. 90

D. 100

E. 104

19) If $y = 5ab + 2b^4$ what is y when $a = 3$ and $= 2$?

A. 34

B. 35

C. 56

D. 62

E. 105

20) The area of a circle is 49π. What is the circumference of the circle?

A. 8π

B. 14π

C. 16π

D. 49π

E. 64π

21) If 80% of x equal to 40% of 20, then what is the value of $(x + 7)^2$?

A. 27.27

B. 27

C. 29.01

D. 289

E. 12,026

22) What is the greatest common factor of 18 and 24?

A. 24

B. 18

C. 10

D. 6

E. 4

23) If the interior angles of a quadrilateral are in the ratio $1:2:3:4$, what is the measure of the smallest angle?

A. $36°$

B. $72°$

C. $108°$

D. $144°$

E. $180°$

24) The length of a rectangle is $\frac{4}{5}$ times its width. If the width is 30, what is the perimeter of this rectangle?

A. 26

B. 58

C. 72

D. 92

E. 108

25) Find $\frac{2}{3}$ of $\frac{3}{4}$ of 160?

A. 80

B. 40

C. 20

D. 4

E. 2

IF YOU FINISH BEFORE TIME IS CALLED, YOU MAY CHECK YOUR WORK ON THIS SECTION ONLY. DO NOT TURN TO ANY OTHER SECTION IN THE TEST.   STOP

SSAT Upper Level Math Practice Test 2

2021

Two Parts

Total number of questions: 50

Section 1: 25 questions

Section 2: 25 questions

Total time for two parts: 60 Minutes

SSAT Upper Level Math Practice Test 2 Answer Sheet

Remove (or photocopy) this answer sheet and use it to complete the practice test.

SSAT Upper Level Mathematics Practice Test 2 Answer Sheet

SSAT Upper Level Math Section 1

1	Ⓐ Ⓑ Ⓒ Ⓓ Ⓔ	11 Ⓐ Ⓑ Ⓒ Ⓓ Ⓔ	21 Ⓐ Ⓑ Ⓒ Ⓓ Ⓔ
2	Ⓐ Ⓑ Ⓒ Ⓓ Ⓔ	12 Ⓐ Ⓑ Ⓒ Ⓓ Ⓔ	22 Ⓐ Ⓑ Ⓒ Ⓓ Ⓔ
3	Ⓐ Ⓑ Ⓒ Ⓓ Ⓔ	13 Ⓐ Ⓑ Ⓒ Ⓓ Ⓔ	23 Ⓐ Ⓑ Ⓒ Ⓓ Ⓔ
4	Ⓐ Ⓑ Ⓒ Ⓓ Ⓔ	14 Ⓐ Ⓑ Ⓒ Ⓓ Ⓔ	24 Ⓐ Ⓑ Ⓒ Ⓓ Ⓔ
5	Ⓐ Ⓑ Ⓒ Ⓓ Ⓔ	15 Ⓐ Ⓑ Ⓒ Ⓓ Ⓔ	25 Ⓐ Ⓑ Ⓒ Ⓓ Ⓔ
6	Ⓐ Ⓑ Ⓒ Ⓓ Ⓔ	16 Ⓐ Ⓑ Ⓒ Ⓓ Ⓔ	
7	Ⓐ Ⓑ Ⓒ Ⓓ Ⓔ	17 Ⓐ Ⓑ Ⓒ Ⓓ Ⓔ	
8	Ⓐ Ⓑ Ⓒ Ⓓ Ⓔ	18 Ⓐ Ⓑ Ⓒ Ⓓ Ⓔ	
9	Ⓐ Ⓑ Ⓒ Ⓓ Ⓔ	19 Ⓐ Ⓑ Ⓒ Ⓓ Ⓔ	
10	Ⓐ Ⓑ Ⓒ Ⓓ Ⓔ	20 Ⓐ Ⓑ Ⓒ Ⓓ Ⓔ	

SSAT Upper Level Math Section 2

1	Ⓐ Ⓑ Ⓒ Ⓓ Ⓔ	11 Ⓐ Ⓑ Ⓒ Ⓓ Ⓔ	21 Ⓐ Ⓑ Ⓒ Ⓓ Ⓔ
2	Ⓐ Ⓑ Ⓒ Ⓓ Ⓔ	12 Ⓐ Ⓑ Ⓒ Ⓓ Ⓔ	22 Ⓐ Ⓑ Ⓒ Ⓓ Ⓔ
3	Ⓐ Ⓑ Ⓒ Ⓓ Ⓔ	13 Ⓐ Ⓑ Ⓒ Ⓓ Ⓔ	23 Ⓐ Ⓑ Ⓒ Ⓓ Ⓔ
4	Ⓐ Ⓑ Ⓒ Ⓓ Ⓔ	14 Ⓐ Ⓑ Ⓒ Ⓓ Ⓔ	24 Ⓐ Ⓑ Ⓒ Ⓓ Ⓔ
5	Ⓐ Ⓑ Ⓒ Ⓓ Ⓔ	15 Ⓐ Ⓑ Ⓒ Ⓓ Ⓔ	25 Ⓐ Ⓑ Ⓒ Ⓓ Ⓔ
6	Ⓐ Ⓑ Ⓒ Ⓓ Ⓔ	16 Ⓐ Ⓑ Ⓒ Ⓓ Ⓔ	
7	Ⓐ Ⓑ Ⓒ Ⓓ Ⓔ	17 Ⓐ Ⓑ Ⓒ Ⓓ Ⓔ	
8	Ⓐ Ⓑ Ⓒ Ⓓ Ⓔ	18 Ⓐ Ⓑ Ⓒ Ⓓ Ⓔ	
9	Ⓐ Ⓑ Ⓒ Ⓓ Ⓔ	19 Ⓐ Ⓑ Ⓒ Ⓓ Ⓔ	
10	Ⓐ Ⓑ Ⓒ Ⓓ Ⓔ	20 Ⓐ Ⓑ Ⓒ Ⓓ Ⓔ	

SSAT Upper Level Math
Practice Test 2

Section 1

25 questions

Total time for this section: 30 Minutes

You may NOT use a calculator for this test.

1) A shaft rotates 200 times in 8 seconds. How many times does it rotate in 12 seconds?

A. 300

B. 250

C. 200

D. 150

E. 100

2) A school wants to give each of its 23 top students a football ball. If the balls are in boxes of three, how many boxes of balls they need to purchase?

A. 3

B. 5

C. 7

D. 8

E. 20

3) If $\frac{25}{A} + 1 = 6$, then $30 + A = ?$

A. 6

B. 1

C. 25

D. 35

E. 0

4) How many tiles of $8 \ cm^2$ is needed to cover a floor of dimension $7 \ cm$ by $24 \ cm$?

A. 6

B. 12

C. 21

D. 24

E. 36

5) Which of the following statements is correct, according to the graph below?

Number of Books Sold in a Bookstore

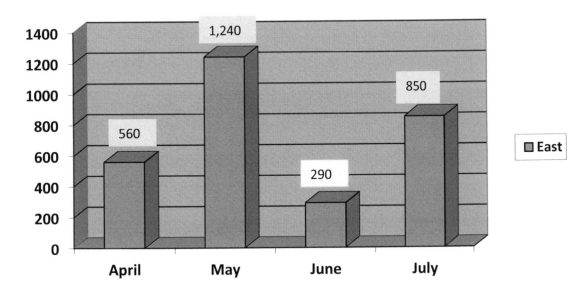

A. Number of books sold in April was twice the number of books sold in July.

B. Number of books sold in July was less than half the number of books sold in May.

C. Number of books sold in June was half the number of books sold in April.

D. Number of books sold in July was equal to the number of books sold in April plus the number of books sold in June.

E. More books were sold in April than in July.

6) $12.124 \div 0.004$?

A. 3.0310

B. 30.310

C. 30,310

D. 3,031

E. 300,310

7) What is the value of the sum of the tens and thousandths in number 5,271.19452?

A. 16

B. 15

C. 14

D. 112

E. 11

8) If 150% of a number is 75, then what is the 80% of that number?

A. 40

B. 50

C. 60

D. 70

E. 85

9) Amy and John work in a same company. Last month, both of them received a raise of 20 percent. If Amy earns $30.00 per hour now and John earns $28.80, Amy earned how much more per hour than John before their raises?

A. $8.25

B. $4.25

C. $3.00

D. $2.25

E. $1.00

10) $\dfrac{1\frac{3}{4}+\frac{1}{3}}{2\frac{1}{2}-\frac{15}{8}}$ is approximately equal to.

A. 3.33

B. 3.6

C. 5.67

D. 6.33

E. 6.67

11) If $2 \leq x < 4$, what is the minimum value of the following expression?

$$2x + 1$$

A. 8

B. 5

C. 3

D. 2

E. 1

12) If $x \blacksquare y = \sqrt{x^2 + y}$, what is the value of $6 \blacksquare 13$?

A. $\sqrt{126}$

B. 7

C. 4

D. 3

E. 2

13) The average weight of 18 girls in a class is 55 kg and the average weight of 32 boys in the same class is 62 kg. What is the average weight of all the 50 students in that class?

A. 50 kg

B. 59.48 kg

C. 61.68 kg

D. 61.9 kg

E. 62.20 kg

14) There are three equal tanks of water. If $\frac{2}{5}$ of a tank contains 250 liters of water, what is the capacity of the three tanks of water together?

A. 1,875 liters

B. 1,200 liters

C. 550 liters

D. 380 liters

E. 200 liters

15) What is the answer of $8.5 \div 0.17$?

A. $\frac{1}{50}$

B. $\frac{1}{5}$

C. 5

D. 50

E. 500

16) Two-kilograms apple and two-kilograms orange cost $26.4. If one-kilogram apple costs $4.2 how much does one-kilogram orange cost?

A. $9

B. $6

C. $5.5

D. $5

E. $4

17) What is the value of x in the following figure?

A. 160

B. 145

C. 125

D. 115

E. 105

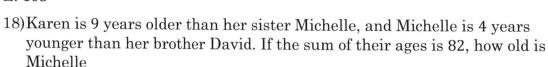

18) Karen is 9 years older than her sister Michelle, and Michelle is 4 years younger than her brother David. If the sum of their ages is 82, how old is Michelle

A. 14

B. 22

C. 23

D. 25

E. 30

19) Michelle and Alec can finish a job together in 50 minutes. If Michelle can do the job by herself in 5 hours, how many minutes does it take Alec to finish the job?

A. 50

B. 60

C. 150

D. 200

E. 220

20) The sum of six different negative integers is -72. If the smallest of these integers is -15, what is the largest possible value of one of the other five integers?

A. -14

B. -10

C. -7

D. -4

E. -1

21) What is the slope of a line that is perpendicular to the line $4x - 2y = 14$?

A. -2

B. $-\frac{1}{2}$

C. 4

D. 12

E. 14

22) A bank is offering 5.5% simple interest on a savings account. If you deposit $7,000, how much interest will you earn in five years?

A. $360

B. $720

C. $1,925

D. $2,600

E. $4,800

23) The Jackson Library is ordering some bookshelves. If x is the number of bookshelves the library wants to order, which each costs \$100 and there is a one-time delivery charge of \$900, which of the following represents the total cost, in dollar, per bookshelf?

A. $100x + 900$

B. $100 + 900x$

C. $\dfrac{100x+900}{100}$

D. $\dfrac{100x+900}{x}$

E. $100x - 900$

24) A football team won exactly 60% of the games it played during last session. Which of the following could be the total number of games the team played last season?

A. 49

B. 35

C. 32

D. 16

E. 12

25) The width of a box is one third of its length. The height of the box is one third of its width. If the length of the box is 36 cm, what is the volume of the box?

A. 81 cm^3

B. 162 cm^3

C. 243 cm^3

D. 729 cm^3

E. 1,728 cm^3

IF YOU FINISH BEFORE TIME IS CALLED, YOU MAY CHECK YOUR WORK ON THIS SECTION ONLY. DO NOT TURN TO ANY OTHER SECTION IN THE TEST.   **STOP**

SSAT Upper Level Math
Practice Test 2

Section 2

25 questions

Total time for this section: 30 Minutes

You may NOT use a calculator for this test.

1) There are 11 marbles in the bag *A* and 19 marbles in the bag *B*. If the sum of the marbles in both bags will be shared equally between two children, how many marbles bag A has less than the marbles that each child will receive?

A. 2

B. 3

C. 4

D. 5

E. 6

2) If the perimeter of the following figure be 26, what is the value of x?

A. 2

B. 3

C. 6

D. 9

E. 12

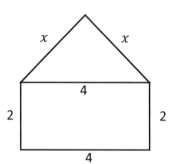

3) When number 92,501 is divided by 305, the result is closest to?

A. 3

B. 30

C. 303

D. 350

E. 400

4) If Jason's mark is k more than Alex, and Jason's mark is 15, which of the following can be Alex's mark?

A. $15 + k$

B. $k - 15$

C. $\frac{k}{15}$

D. $15k$

E. $15 - k$

5) To paint a wall with the area of $62m^2$, how many liters of paint do we need if each liter of paint is enough to paint a wall with dimension of $62\ cm \times 100\ cm$?

A. 100

B. 120

C. 150

D. 200

E. 250

6) The price of a sofa is decreased by 25% to \$432. What was its original price?

A. \$480

B. \$520

C. \$576

D. \$600

E. \$800

7) $750 - 8\frac{7}{15} = ?$

A. $741\frac{7}{15}$

B. $741\frac{8}{15}$

C. $743\frac{1}{15}$

D. $743\frac{8}{15}$

E. $744\frac{1}{15}$

8) A driver rests one hour and 12 minutes for every 3 hours driving. How many minutes will he rest if he drives 15 hours?

A. 3 hours and 36 minutest

B. 4 hours and 12 minutest

C. 4 hours and 45 minutest

D. 5 hours and 36 minutest

E. 6 hours

9) Which of the following expression is not equal to 4?

A. $8 \times \frac{1}{2}$

B. $20 \times \frac{1}{5}$

C. $2 \times \frac{4}{2}$

D. $4 \times \frac{5}{5}$

E. $4 \times \frac{1}{4}$

10) What is the missing term in the given sequence?

$$2, 3, 5, 8, 12, 17, \underline{\quad}, 30$$

A. 23

B. 24

C. 27

D. 28

E. 30

Questions 11 to 12 are based on the following graph

A library has 800 books that include Mathematics, Physics, Chemistry, English and History.

Use following graph to answer questions 11 to 12.

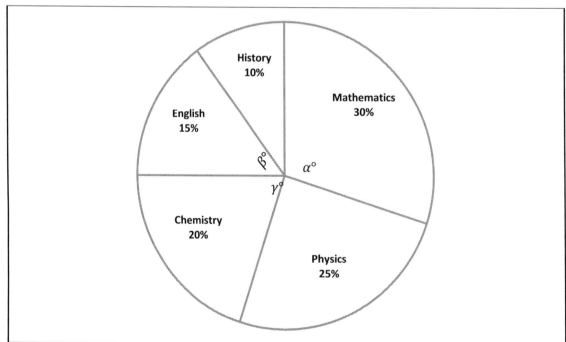

11) What is the product of the number of Mathematics and number of English books?

A. 21,168

B. 28,800

C. 29,460

D. 30,640

E. 35,280

12) What are the values of angle α and β respectively?

A. $90°, 54°$

B. $120°, 36°$

C. $120°, 45°$

D. $108°, 54°$

E. $108°, 45°$

13) The length of a rectangle is 3 times of its width. If the length is 24, what is the perimeter of the rectangle?

A. 24

B. 30

C. 36

D. 48

E. 64

14) If $3y + 2 < 29$, then y could be equal to?

A. 15

B. 12

C. 10.5

D. 9

E. 2.5

15) The capacity of a red box is 20% bigger than the capacity of a blue box. If the red box can hold 36 equal sized books, how many of the same books can the blue box hold?

A. 9

B. 15

C. 21

D. 30

E. 36

16) If the area of the following rectangular $ABCD$ is 120, and E is the midpoint of AB, what is the area of the shaded part?

A. 25

B. 60

C. 75

D. 80

E. 120

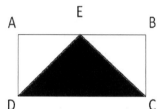

17) There are 60.2 liters of gas in a car fuel tank. In the first week and second week of April, the car uses 5.28 and 25.9 liters of gas respectively. If the car was park in the third week of April and 10.31 liters of gas will be added to the fuel tank, how many liters of gas are in the fuel tank of the car?

A. 21.41 liters

B. 25.9 liters

C. 27 liters

D. 39.33 liters

E. 71.61 liters

18) If $3x + y = 25$ and $x - z = 18$, what is the value of x?

A. 0

B. 5

C. 10

D. 20

E. it cannot be determined from the information given

19) If 96 is the product of 4 and $8x$, then 96 is divisible by which of the following?

A. $x + 4$

B. $2x - 1$

C. $5x - 3$

D. $x \times 3$

E. $3x + 1$

20) What is the average of circumference of figure A and area of figure B? ($\pi = 3$)

A. 50

B. 53

C. 52

D. 51

E. 50

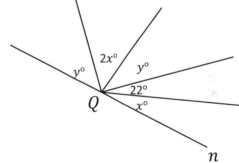

Figure A

Figure B

21) In the following figure, point Q lies on line n, what is the value of y if $x = 28$?

A. 32

B. 37

C. 42

D. 45

E. 56

22) Which of the following is closest to $\frac{1}{6}$ of 40?

A. 0.3×6

B. 0.3×5

C. 0.2×30

D. 0.2×35

E. 0.2×39.5

23) A number is chosen at random from 1 to 18. Find the probability of not selecting a composite number. (A composite number is a number that is divisible by itself, 1 and at least one other whole number)

A. $\frac{1}{18}$

B. $\frac{1}{3}$

C. $\frac{7}{18}$

D. 1

E. 0

24) $812 \div 3 =?$

A. $\frac{800}{3} \times \frac{10}{3} \times \frac{2}{3}$

B. $800 + \frac{10}{3} + \frac{2}{3}$

C. $\frac{800}{3} + \frac{10}{3} + \frac{2}{3}$

D. $\frac{800}{3} \div \frac{10}{3} \div \frac{2}{3}$

E. $\frac{8}{3} + \frac{1}{3} + \frac{2}{3}$

25) If a gas tank can hold 35 gallons, how many gallons does it contain when it is $\frac{2}{5}$ full?

A. 125

B. 62.5

C. 50

D. 14

E. 5

IF YOU FINISH BEFORE TIME IS CALLED, YOU MAY CHECK YOUR WORK ON THIS SECTION ONLY. DO NOT TURN TO ANY OTHER SECTION IN THE TEST. STOP

SSAT Upper Level Math Practice Test Answer Keys

Now, it's time to review your results to see where you went wrong and what areas you need to improve.

SSAT Upper Level Math Practice Test 1							
Section 1				**Section 2**			
1	C	16	A	1	C	16	B
2	A	17	B	2	B	17	D
3	C	18	C	3	E	18	E
4	E	19	B	4	D	19	D
5	B	20	D	5	A	20	B
6	D	21	D	6	D	21	D
7	D	22	B	7	D	22	D
8	C	23	A	8	D	23	A
9	B	24	C	9	E	24	E
10	B	25	E	10	C	25	A
11	B			11	D		
12	B			12	C		
13	A			13	C		
14	C			14	C		
15	A			15	C		

SSAT Upper Level Math Practice Test 2

Section 1				Section 2			
1	A	16	A	1	C	16	B
2	D	17	A	2	D	17	D
3	D	18	C	3	C	18	E
4	C	19	B	4	E	19	C
5	D	20	C	5	A	20	A
6	D	21	B	6	C	21	B
7	E	22	C	7	B	22	D
8	A	23	D	8	E	23	C
9	E	24	B	9	E	24	C
10	A	25	E	10	A	25	D
11	B			11	B		
12	B			12	D		
13	B			13	E		
14	A			14	E		
15	D			15	D		

SSAT Upper Level Math Practice
Test Answers and Explanations

SSAT Upper Level Mathematics Practice Test 1 Section 1

1) Choice C is correct

Digit 9 is in the hundredths place.

2) Choice A is correct

$x - 20 = -20 \rightarrow x = -20 + 20 \rightarrow x = 0$, Then; $x \times 20 = 0 \times 20 = 0$

3) Choice C is correct

$0.04 \times 13.00 = \frac{4}{100} \times \frac{13}{1} = \frac{52}{100} = 0.52$

4) Choice E is correct

His average speed was: $\frac{3.5}{0.5} = 7$ miles per hour

5) Choice B is correct

The perimeter of the quadrilateral is: $7 + 21 + 10 + 32 = 70$

6) Choice D is correct

The amount of flour is: $M - 75$

7) Choice D is correct

Number of packs needed equals to: $\frac{18}{4} \cong 4.5$, Then Mia must purchase 5 packs.

8) Choice C is correct

The time it takes to drive from city A to city B is: $\frac{3,600}{70} = 51.42$, It's approximately 51 hours.

9) Choice B is correct

Number of males in the classroom is: $60 - 30 = 30$

Then, the percentage of males in the classroom is: $\frac{30}{60} \times 100 = 0.5 \times 100 = 50\%$

10) Choice B is correct

For one person the total cost is: $9.20, Therefore, for five persons, the total cost is: $5 \times \$9.20 = \$46$

11) Choice B is correct

Let x be the fourth quarter rate, then: $\frac{93+40+87+x}{4} = 97$, Multiply both sides of the above equation by 4. Then: $4 \times \left(\frac{93+40+87+x}{4}\right) = 4 \times 97 \rightarrow 93 + 40 + 87 + x = 388 \rightarrow 220 + x = 388 , \rightarrow x = 388 - 220 = 168$

12) Choice B is correct

Let x be the number. Write the equation and solve for x. $\frac{2}{3} \times 15 = 10 \rightarrow 10 = \frac{2}{5}x$, multiply both sides of the equation by $\frac{5}{2}$, then: $10 \times \frac{5}{2} = \frac{2}{5}x \times \frac{5}{2} \rightarrow x = 25$

13) Choice A is correct

Use the information provided in the question to draw the shape.

Use Pythagorean Theorem: $a^2 + b^2 = c^2$

$60^2 + 80^2 = c^2 \Rightarrow 3,600 + 6,400 = c^2 \Rightarrow 10,000 = c^2 \Rightarrow c = 100 \; miles$

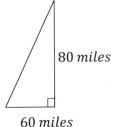

80 *miles*

Port A

60 *miles*

14) Choice C is correct

Three times of 25,000 is 75,000. One sixth of them cancelled their tickets.

One sixth of 75,000 equals 12,500 ($\frac{1}{6} \times 75,000 = 12,500$).

62,500 ($75,000 - 12,500 = 62,500$) fans are attending this week

15) Choice A is correct

The equation of a line in slope intercept form is: $y = mx + b$, Solve for y: $8x + y = 14 \Rightarrow y = -8x + 14$, The slope of this line is -8. The slope of the line perpendicular to this line is: $m_1 \times m_2 = -1 \Rightarrow -8 \times m_2 = -1 \Rightarrow m_2 = \frac{1}{8}$

16) Choice A is correct

Let x be the number of years. Therefore, $3,000 per year equals $3,000x$. Starting from $34,000 annual salary means you should add that amount to $3,000x$. Income more than that is: $I > 3,000x + 34,000$

17) Choice B is correct

The length of MN is equal to: $4x + 6x = 10x$, Then: $10x = 50 \rightarrow x = \frac{50}{10} = 5$

The length of ON is equal to: $6x = 6 \times 5 = 30 \; cm$

18) Choice C is correct

Let the number of women should be added to city D be x, then:

$\frac{528 + x}{550} = 1.2 \rightarrow 528 + x = 550 \times 1.2 = 660 \rightarrow x = 132$

19) Choice B is correct

Ratio of women to men in city A: $\frac{570}{600} = 0.95$

Ratio of women to men in city B: $\frac{291}{300} = 0.97$

Ratio of women to men in city C: $\frac{665}{700} = 0.95$

Ratio of women to men in city D: $\frac{528}{550} = 0.96$

0.97 is the maximum ratio of woman to man in the four cities.

20) Choice D is correct

Percentage of men in city $= \frac{600}{1,170} \times 100 = 51.28\%$, Percentage of women in city $C = \frac{665}{1,365} \times 100 = 48.72\%$, Percentage of men in city A to percentage of women in city $C = \frac{51.28}{48.72} = 1.05$

21) Choice D is correct

Smallest 5–digit number is 10,000, and biggest 5–digit number is 99,999. The difference is: 89,999

22) Choice B is correct

$\frac{x}{3+4} = \frac{y}{11-8} \rightarrow \frac{x}{7} = \frac{y}{3} \rightarrow 7y = 3x \rightarrow y = \frac{3}{7}x$

23) Choice A is correct

$\sqrt[5]{x^{21}} = \sqrt[5]{x^{20} \times x} = \sqrt[5]{x^{20}} \times \sqrt[5]{x} = x^{\frac{20}{5}} \times \sqrt[5]{x} = x^4\sqrt[5]{x}$

24) Choice C is correct

Perimeter of rectangle A is equal to: $2 \times (10 + 6) = 2 \times 16 = 32$

Perimeter of rectangle B is equal to: $2 \times (6 + 4) = 2 \times 10 = 20$

Therefore: $\frac{20}{32} \times 100 = 0.625 \times 100 = 62.5\% \cong 62\%$

25) Choice E is correct

The average speed of John is: $160 \div 4 = 40 \ km$, The average speed of Alice is: $190 \div 5 = 38 \ km$, Write the ratio and simplify. $40 : 38 \Rightarrow 20 : 19$

SSAT Upper Level Mathematics Practice Test 1 Section 2

1) Choice C is correct

$0.44 \times 12.8 = \frac{44}{100} \times \frac{128}{10} = \frac{44 \times 128}{100 \times 10} = \frac{5,632}{1,000} = 5.632$

2) Choice B is correct

$4\frac{3}{8} \times 5\frac{1}{9} = \frac{35}{8} \times \frac{46}{9} = \frac{35 \times 46}{8 \times 9} = \frac{1,610}{72} = \frac{805}{36} = 22\frac{13}{36}$

3) Choice E is correct

A. $\frac{3}{2} \times \frac{5}{9} = \frac{5}{6}$ is not a whole number

B. $\frac{1}{3} + \frac{1}{4} = \frac{4+3}{12} = \frac{7}{12}$ is not a whole number

C. $\frac{40}{6} = \frac{20}{3} = 6.66$ is not a whole number

D. $3.5 + 2 = 5.5$ is not a whole number

E. $3.5 + \frac{4}{8} = 3.5 + 0.5 = 4$ is a whole number

4) Choice D is correct

8 cubed is: $8 \times 8 \times 8 = 64 \times 8 = 512$

5) Choice A is correct

Let x be the number, then; $\frac{2}{3}x = 16 \rightarrow x = \frac{3 \times 16}{2} = 24$, Therefore: $\frac{5}{3}x = \frac{5}{3} \times 24 = 40$

6) Choice D is correct

Use formula of rectangle prism volume. $V = (length)(width)(height) \Rightarrow 3,000 = (30)(10)(height). \Rightarrow height = 3,000 \div 300 = 10$ feet

7) Choice D is correct

Number of boxes equal to: $\frac{450}{25} = \frac{90}{5} = 18$

8) Choice D is correct

The angle x and 60 are complementary angles. Therefore: $x + 60 = 180$, $180° - 60° = 120°$

9) Choice E is correct

Use simple interest formula: $I = prt$ ($I =$ interest, $p =$ principal, $r =$ rate, $t =$ time)

$I = (13,500)(0.045)(2) = \$1,215$

10) Choice C is correct

$\frac{504}{600} = 0.84 = 84\%$. 504 is 84% of 600. Therefore, the discount is: $100\% - 84\% = 16\%$

11) Choice D is correct

Use distance formula: $Distance = Rate \times time \Rightarrow 527 = 62 \times T$, divide both sides by 62.

$\frac{527}{62} = T \Rightarrow T = 8.5 \ hours$. Change hours to minutes for the decimal part.

$0.5 \ hours = 0.5 \times 60 = 30 \ minutes$

12) Choice C is correct

The rate of construction company $= \frac{40 \ cm}{1 \ min} = 40 \frac{cm}{min}$

Height of the wall after $50 \ min = \frac{40 \ cm}{1 \ min} \times 50 \ min = 2,000 cm$

Let x be the height of wall, then $\frac{2}{3}x = 2,000 \ cm \rightarrow x = \frac{3 \times 2,000}{2} \rightarrow x = 3,000 \ cm = 30m$

13) Choice C is correct

Let's review the choices provided:

A. $28 - 2 = 26 \rightarrow \frac{26}{2} = 13 \neq 3$

B. $28 - 5 = 23 \rightarrow \frac{23}{5} = 4.6 \neq 3$

C. $28 - 7 = 21 \rightarrow \frac{21}{7} = 3 = 3$

D. $28 - 12 = 16 \rightarrow \frac{16}{12} = 1.33 \neq 3$

E. $28 - 28 = 0 \rightarrow \frac{0}{28} = 0 \neq 3$

14) Choice C is correct

Speed of car A is: $\frac{700}{7} = 100$ miles per hour , Speed of car B is: $\frac{700}{5} = 140$ miles per hour

$\rightarrow 140 - 100 = 40$ miles per hour

15) Choice C is correct

x is the number of all sales profit and 4% of it is: $4\% \times x = 0.04x$, Employee's revenue: $0.04x + 8000$

16) Choice B is correct

$\frac{1}{5} \cong 0.2$ $\qquad$ $\frac{5}{3} \cong 1.66 \, \frac{8}{11} \cong 0.73$ $\qquad$ $\frac{2}{3} = 0.66$

$\frac{5}{3} > \frac{8}{11} > \frac{2}{3} > \frac{1}{5}$

17) Choice D is correct

The ratio of boys to girls is $3 : 2$. Therefore, there are 3 boys out of 5 students. To find the answer, first divide the total number of students by 5, then multiply the result by 3.

$700 \div 5 = 140 \Rightarrow 140 \times 3 = 420$

18) Choice E is correct

$\frac{(8+6)^2}{2} + 6 = \frac{(14)^2}{2} + 6 = \frac{196}{2} + 6 = 98 + 6 = 104$

19) Choice D is correct

$y = 5ab + 2b^4$, $\qquad$ Plug in the values of a and b in the equation: $a = 3$ and $b = 2$

$y = 5\,(3)(2) + 2(2)^4 = 30 + 2(16) = 30 + 32 = 62$

20) Choice B is correct

Use the formula of the area of circles. $Area = \pi r^2 \Rightarrow 49\pi = \pi r^2 \Rightarrow 49 = r^2 \Rightarrow r = 7$

Radius of the circle is 7. Now, use the circumference formula: Circumference $= 2\pi r = 2\pi\,(7) = 14\pi$

21) Choice D is correct

$0.8x = (0.4) \times 20 \rightarrow x = 10 \rightarrow (x + 7)^2 = (10 + 7)^2 = (17)^2 = 289$

22) Choice D is correct

Prime factorizing of $18 = 2 \times 3 \times 3$, Prime factorizing of $24 = 2 \times 2 \times 2 \times 3$

GCF $= 2 \times 3 = 6$

23) Choice A is correct

The sum of all angles in a quadrilateral is 360 degrees. Let x be the smallest angle in the quadrilateral. Then the angles are: $x, 2x, 3x, 4x$,

$x + 2x + 3x + 4x = 360 \rightarrow 10x = 360 \rightarrow x = 36$, The angles in the quadrilateral are: $36°, 72°, 108°$, and $144°$, The smallest angle is 36 degrees.

24) Choice E is correct

Length of the rectangle is: $\frac{4}{5} \times 30 = 24$, Perimeter of rectangle is: $2 \times (24 + 30) = 108$

25) Choice A is correct

$\frac{3}{4}$ of $160 = \frac{3}{4} \times 160 = 120$, $\frac{2}{3}$ of $120 = \frac{2}{3} \times 120 = 80$

SSAT Upper Level Mathematics Practice Test 2 Section 1

1) Choice A is correct

Number of rotates in 12 second equals to: $\frac{200 \times 12}{8} = 300$

2) Choice D is correct

Number of packs equal to: $\frac{23}{3} \cong 7.666$, Therefore, the school must purchase 8 packs.

3) Choice D is correct

$\frac{25}{A} + 1 = 6 \rightarrow \frac{25}{A} = 6 - 1 = 5, \rightarrow 25 = 5A \rightarrow A = \frac{25}{5} = 5,\ 30 + A = 30 + 5 = 35$

4) Choice C is correct

The area of the floor is: $7\ cm \times 24\ cm = 168\ cm^2$, The number of tiles needed $=$

$168 \div 8 = 21$

5) Choice D is correct

Number of books sold in April is: 560

Number of books sold in July is: $850 \rightarrow \frac{560}{850} = \frac{56}{85} \cong 0.66$

number of books sold in July is: 850

Half the number of books sold in May is: $\frac{1,240}{2} = 620 \rightarrow 850 > 620$

number of books sold in June is: 290

Half the number of books sold in April is: $\frac{560}{2} = 280 \rightarrow 290 \neq 280$

$560 + 290 = 850 = 850$

$560 < 850$

6) Choice D is correct

$12.124 \div 0.004 = \dfrac{\frac{12,124}{1,000}}{\frac{4}{1,000}} = \dfrac{12,124}{4} = 3,031$

7) Choice E is correct

The digit in tens place is 7. The digit in the thousandths place is 4. Therefore; $7 + 4 = 11$

8) Choice A is correct

First, find the number. Let x be the number. Write the equation and solve for x. 150% of a number is 75, then: $1.5 \times x = 75 \rightarrow x = 75 \div 1.5 = 50$, 80% of 50 is: $0.8 \times 50 = 40$

9) Choice E is correct

Amy earns $30.00 *per hour* now. $30.00 *per hour* is 20 percent more than her previous rate. Let x be her rate before her raise. Then: $x + 0.20x = 30 \rightarrow 1.2x = 30 \rightarrow x = \frac{30}{1.2} = 25$

John earns $28.80 *per hour* now. $28.80 *per hour* is 20 percent more than his previous rate. Let x be John's rate before his raise. Then: $x + 0.20x = 28.80 \rightarrow 1.2x = 28.80 \rightarrow x = \frac{28.80}{1.2} = 24$, Amy earned $1.00 more per hour than John before their raises

10) Choice A is correct

$$\frac{1\frac{3}{4}+\frac{1}{3}}{2\frac{1}{2}-\frac{15}{8}} = \frac{\frac{7}{4}+\frac{1}{3}}{\frac{5}{2}-\frac{15}{8}} = \frac{\frac{21+4}{12}}{\frac{20-15}{8}} = \frac{\frac{25}{12}}{\frac{5}{8}} = \frac{25 \times 8}{12 \times 5} = \frac{5 \times 2}{3 \times 1} = \frac{10}{3} \cong 3.33$$

11) Choice B is correct

$2 \leq x < 4 \rightarrow$ Multiply all sides of the inequality by 2. Then: $2 \times 2 \leq 2 \times x < 2 \times 4 \rightarrow 4 \leq 2x < 8$, All 1 to all sides. Then: $\rightarrow 4 + 1 \leq 2x + 1 < 8 + 1 \rightarrow 5 \leq 2x + 1 < 9$

Minimum value of $2x + 1$ is 5.

12) Choice B is correct

$6 \blacksquare 13 = \sqrt{6^2 + 13} = \sqrt{36 + 13} = \sqrt{49} = 7$

13) Choice B is correct

$Average = \frac{sum\ of\ terms}{number\ of\ terms}$, The sum of the weight of all girls is: $18 \times 55 = 990\ kg$

The sum of the weight of all boys is: $32 \times 62 = 1,984\ kg$, The sum of the weight of all students is: $990 + 1,984 = 2,974 kg$, The average weight of the 50 students: $Average = \frac{2,974}{50} = 59.48\ kg$

14) Choice A is correct

Let x be the capacity of one tank. Then, $\frac{2}{5}x = 250 \rightarrow x = \frac{250 \times 5}{2} = 625$ Liters

The amount of water in three tanks is equal to: $3 \times 625 = 1,875$ Liters

15) Choice D is correct

$$8.5 \div 0.17 = \frac{8.5}{0.17} = \frac{\frac{85}{10}}{\frac{17}{100}} = \frac{85 \times 100}{17 \times 10} = \frac{85}{17} \times \frac{100}{10} = 5 \times 10 = 50$$

16) Choice A is correct

Let x be the cost of one-kilogram orange, then: $2x + (2 \times 4.2) = 26.4 \rightarrow 2x + 8.4 = 26.4 \rightarrow 2x = 26.4 - 8.4 \rightarrow 2x = 18 \rightarrow x = \frac{18}{2} = \$9$

17) Choice A is correct

$x = 35 + 125 = 160$

18) Choice C is correct

Let's write equations based on the information provided:

$Michelle = Karen - 9, Michelle = David - 4, Karen + Michelle + David = 82$

$Karen - 9 = Michelle \Rightarrow Karen = Michelle + 9$

$Karen + Michelle + David = 82$

Now, replace the ages of Karen and David by Michelle. Then:

$Michelle + 9 + Michelle + Michelle + 4 = 82$

$3 \times Michelle + 13 = 82 \Rightarrow 3 \times Michelle = 82 - 13$

$3 \times Michelle = 69 , Michelle = 23$

19) Choice B is correct

Let b be the amount of time Alec can do the job, then,

$\frac{1}{a} + \frac{1}{b} = \frac{1}{50} \rightarrow \frac{1}{300} + \frac{1}{b} = \frac{1}{50} \rightarrow \frac{1}{b} = \frac{1}{50} - \frac{1}{300} = \frac{5}{300} = \frac{1}{60}$, Then: $b = 60$ minutes

20) Choice C is correct

The smallest number is -15. To find the largest possible value of one of the other five integers, we need to choose the smallest possible integers for four of them. Let x be the largest number. Then: $-72 = (-15) + (-14) + (-13) + (-12) + (-11) + x \rightarrow -72 = -65 + x, \rightarrow x = -72 + 65 = -7$

21) Choice B is correct

The equation of a line in slope intercept form is: $y = mx + b$. Solve for y. $4x - 2y = 14 \Rightarrow -2y = 14 - 4x \Rightarrow y = (14 - 4x) \div (-2) \Rightarrow y = 2x - 7$

The slope is 2. The slope of the line perpendicular to this line is: $m_1 \times m_2 = -1 \Rightarrow 2 \times m_2 = -1 \Rightarrow m_2 = -\frac{1}{2}$

22) Choice C is correct

Use simple interest formula: $I = prt$ (I = interest, p = principal, r = rate, t = time)

$I = (7,000)(0.055)(5) = \$1,925$

23) Choice D is correct

The amount of money for x bookshelf is: $100x$, Then, the total cost of all bookshelves is equal to: $100x + 900$, The total cost, in dollar, per bookshelf is: $\frac{\text{Total cost}}{\text{number of items}} = \frac{100x+900}{x}$

24) Choice B is correct

Choices A, C and D are incorrect because 60% of each of the numbers is non-whole number.

A. 49,   60% of $49 = 0.60 \times 49 = 29.4$

B. 35,   60% of $35 = 0.60 \times 35 = 21$

C. 32,   60% of $32 = 0.60 \times 32 = 19.2$

D. 16,   60% of $16 = 0.60 \times 16 = 9.6$

E. 12,   60% of $12 = 0.60 \times 12 = 7.2$

25) Choice E is correct

If the length of the box is 36, then the width of the box is one third of it, 12, and the height of the box is 4 (one third of the width). The volume of the box is:

$V = (length)(width)(height) = (36)(12)(4) = 1,728 \ cm^3$

SSAT Upper Level Mathematics Practice Test 2 Section 2

1) Choice C is correct

$\frac{19+11}{2} = \frac{30}{2} = 15$ Then, $15 - 11 = 4$

2) Choice D is correct

Let's review the choices provided:

A. $x = 2 \rightarrow$ The perimeter of the figure is: $2 + 4 + 2 + 2 + 2 = 12 \neq 26$

B. $x = 3 \rightarrow$ The perimeter of the figure is: $2 + 4 + 2 + 3 + 3 = 14 \neq 26$

C. $x = 6 \rightarrow$ The perimeter of the figure is: $2 + 4 + 2 + 6 + 6 = 20 \neq 26$

D. $x = 9 \rightarrow$ The perimeter of the figure is: $2 + 4 + 2 + 9 + 9 = 26 = 26$

E. $x = 12 \rightarrow$ The perimeter of the figure is: $2 + 4 + 2 + 12 + 12 = 32 \neq 26$

3) Choice C is correct

$\frac{92,501}{305} \cong 303.2819 \cong 303$

4) Choice E is correct

Alex's mark is k less than Jason's mark. Then, from the choices provided Alex's mark can only be $15 - k$.

5) Choice A is correct

The Area that one liter of paint is required: $62cm \times 100cm = 6,200cm^2$

Remember: $1\,m^2 = 10,000\,cm^2\,(100 \times 100 = 10,000), then, 6,200cm^2 = 0.62\,m^2$

Number of liters of paint we need: $\frac{62}{0.62} = 100$ liters

6) Choice C is correct

Let x be the original price. If the price of the sofa is decreased by 25% to \$432, then: $75\,\%\,of\,x = 432 \Rightarrow 0.75x = 432 \Rightarrow x = 432 \div 0.75 = 576$

7) Choice B is correct

$750 - 8\frac{7}{15} = (749 - 8) + \left(\frac{15}{15} - \frac{7}{15}\right) = 741\frac{8}{15}$

8) Choice E is correct

Number of times that the driver rests $= \frac{15}{3} = 5$, Driver's rest time $=$ 1 *hour and* 12 *minutes* $= 72$ *minutes*, Then, 5×72 minutes $= 360$ minutes, 1 *hour* $= 60$ *minutes* $\rightarrow 360$ *minutes* $= 6$ *hours*

9) Choice E is correct

Let's review the options provided:

A. $8 \times \frac{1}{2} = \frac{8}{2} = 4 = 4$

B. $20 \times \frac{1}{5} = \frac{20}{5} = 4 = 4$

C. $2 \times \frac{4}{2} = \frac{8}{2} = 4 = 4$

D. $4 \times \frac{5}{5} = \frac{20}{5} = 4 = 4$

E. $4 \times \frac{1}{4} = \frac{4}{4} = 1 \neq 4$

10) Choice A is correct

Find the difference of each pairs of numbers: $2, 3, 5, 8, 12, 17, \_\_, 30$

The difference of 2 and 3 is 1, 3 and 5 is 2, 5 and 8 is 3, 8 and 12 is 4, 12 and 17 is 5, 17 and next number should be 6. The number is $17 + 6 = 23$

11) Choice B is correct

Number of Mathematics book: $0.3 \times 800 = 240$, Number of English books: $0.15 \times 800 = 120$, Product of number of Mathematics and number of English books: $240 \times 120 = 28,800$

12) Choice D is correct

The angle α is: $0.3 \times 360 = 108°$, The angle β is: $0.15 \times 360 = 54°$

13) Choice E is correct

The length of the rectangle is 24. Then, its width is 8. $24 \div 3 = 8$

$$Perimeter\ of\ a\ rectangle = 2 \times width + 2 \times length = 2 \times 8 + 2 \times 24 = 16 + 48$$
$$= 64$$

14) Choice E is correct

$3y + 2 < 29 \rightarrow 3y < 29 - 2 \rightarrow 3y < 27 \rightarrow y < 9$, The only choice that is less than 9 is E.

15)Choice D is correct

The capacity of a red box is 20% bigger than the capacity of a blue box and it can hold 36 books. Therefore, we want to find a number that 20% bigger than that number is 36. Let x be that number. Then: $1.20 \times x = 36$, Divide both sides of the equation by 1.2. Then:

$$x = \frac{36}{1.20} = 30$$

16)Choice B is correct

Since, E is the midpoint of AB, then the area of all triangles DAE, DEF, CFE and CBE are equal. Let x be the area of one of the triangle, then:

$$4x = 120 \rightarrow x = 30$$

The area of $DEC = 2x = 2(30) = 60$

17)Choice D is correct

Amount of available petrol in tank: $60.2 - 5.28 - 25.9 + 10.31 = 39.33$ liters

18)Choice E is correct

We have two equations and three unknown variables, therefore x cannot be obtained.

19)Choice C is correct

$96 = 8x \times 4 \rightarrow x = 96 \div 4 = 24 \rightarrow x = 3$

x equals to 3. Let's review the choices provided:

A. $x + 4 \rightarrow 3 + 4 = 7 \qquad$ 96 is not divisible by 7.

B. $2x - 1 \rightarrow 2 \times 3 - 1 = 5 \quad$ 96 is not divisible by 5.

C. $5x - 3 \rightarrow 5 \times 3 - 3 = 12$ 96 is divisible by 12.

D. $x \times 3 \rightarrow 3 \times 3 = 9 \qquad$ 96 is not divisible by 9.

E. $3x + 1 \rightarrow 3 \times 3 + 1 = 10$ 96 is not divisible by 10.

The answer is C

20)Choice A is correct

Perimeter of figure A is: $2\pi r = 2\pi \frac{16}{2} = 16\pi = 16 \times 3 = 48$, Area of figure B is: $4 \times 13 = 52$, $Average = \frac{48+52}{2} = \frac{100}{2} = 50$

21) Choice B is correct

The angles on a straight line add up to 180 degrees. Then: $x + 22 + y + 2x + y = 180$, Then, $3x + 2y = 180 - 22 \rightarrow 3(28) + 2y = 158 \rightarrow 2y = 158 - 84 = 74 \rightarrow y = 37$

22) Choice D is correct

$\frac{1}{6}$ of 40 is 6.66. Let's review the choices provided:

A. $0.3 \times 6 = 1.8$

B. $0.3 \times 5 = 1.5$

C. $0.2 \times 30 = 6$

D. $0.2 \times 35 = 7$

E. $0.2 \times 39.5 = 7.9$

Option D is the closest to 6.66

23) Choice C is correct

Set of numbers that are not composite between 1 and 18: $A = \{2, 3, 5, 7, 11, 13, 17\}$

$$Probability = \frac{number\ of\ desired\ outcomes}{number\ of\ total\ outcomes} = \frac{7}{18}$$

24) Choice C is correct

$$812 \div 3 = \frac{812}{3} = \frac{800 + 10 + 2}{3} = \frac{800}{3} + \frac{10}{3} + \frac{2}{3}$$

25) Choice D is correct

$\frac{2}{5} \times 35 = \frac{70}{5} = 14$

... So Much More Online!

Effortless Math Online SSAT Upper Level Center offers a complete study program, including the following:

- ✓ Step-by-step instructions on how to prepare for the SSAT Upper Level test

- ✓ Numerous SSAT Upper Level worksheets to help you measure your math skills

- ✓ Complete list of SSAT formulas

- ✓ Video lessons for SSAT topics

- ✓ Full-length SSAT practice tests

- ✓ And much more...

No Registration Required.

Receive the PDF version of this book or get another FREE book!

Thank you for using our Book!

Do you LOVE this book?

Then, you can get the PDF version of this book or another book absolutely FREE!

Please email us at:

info@EffortlessMath.com

for details.

Author's Final Note

I hope you enjoyed reading this book. You've made it through the book! Great job!

First of all, thank you for purchasing this practice book. I know you could have picked any number of books to help you prepare for your SSAT Upper Level test, but you picked this book and for that I am extremely grateful.

It took me years to write this workbook for the SSAT Upper Level because I wanted to prepare a comprehensive SSAT Upper Level workbook to help test takers make the most effective use of their valuable time while preparing for the test.

After teaching and tutoring math courses for over a decade, I've gathered my personal notes and lessons to develop this practice book. It is my greatest hope that the exercises in this book could help you prepare for your test successfully.

If you have any questions, please contact me at reza@effortlessmath.com and I will be glad to assist. Your feedback will help me to greatly improve the quality of my books in the future and make this book even better. Furthermore, I expect that I have made a few minor errors somewhere in this book. If you think this to be the case, please let me know so I can fix the issue as soon as possible.

If you enjoyed this book and found some benefit in reading this, I'd like to hear from you and hope that you could take a quick minute to post a review on the book's Amazon page. To leave your valuable feedback, please visit: amzn.to/39VImZb

Or scan this QR code.

I personally go over every single review, to make sure my books really are reaching out and helping students and test takers. Please help me help SSAT Upper Level test takers, by leaving a review!

I wish you all the best in your future success!

Reza Nazari

Math teacher and author

Made in the USA
Middletown, DE
25 August 2021